21st Century Patton

TITLES IN THE SERIES

21st Century Mahan
21st Century Sims
21st Century Ellis
21st Century Knox

21ST CENTURY FOUNDATIONS

Benjamin Armstrong, *editor*

In 1911 Capt. Alfred Thayer Mahan wrote in his book *Armaments and Arbitration*, "The study of military history lies at the foundation of all sound military conclusions and practices." One hundred years later, as we sail ever further into the twenty-first century, we are commonly told we face the most challenging circumstances in history, or that it is more dangerous now than ever before. These exaggerations tend to ignore the lessons of strategy and policy that come from our past.

The 21st Century Foundations series gives modern perspective to the great strategists and military philosophers of the past, placing their writings, principles, and theories within modern discussions and debates. Whether drawn from famous men or more obscure contributors with lesser known works, collecting and analyzing their writing will inform a new generation of students, military professionals, and policy makers alike. The essays and papers collected in this series are not provided in order to spell out cut and dry answers or exact procedures, but instead to help make sure we ask the right questions as we face the challenges of the future. The series informs the present by collecting and offering strategists and thinkers of the past.

21st Century Patton

Strategic Insights for the Modern Era

EDITED BY J. FURMAN DANIEL III

Naval Institute Press
Annapolis, Maryland

This book has been brought to publication with the generous assistance of Marguerite and Gerry Lenfest.

Naval Institute Press
291 Wood Road
Annapolis, MD 21402

Library of Congress Cataloging-in-Publication Data

Names: Daniel, J. Furman, III, author. | Patton, George S. (George Smith), 1885–1945, author.
Title: 21st century Patton : strategic insights for the modern era / edited by J. Furman Daniel III.
Other titles: Twenty-first century Patton, strategic insights for the modern era
Description: Annapolis, Maryland : Naval Institute Press, [2016] | Series: 21st century foundations | Includes bibliographical references and index.
Identifiers: LCCN 2016021668| ISBN 9781682470633 (alk. paper) | ISBN 9781682470640 (mobi)
Subjects: LCSH: Patton, George S. (George Smith), 1885–1945—Military leadership. | Patton, George S. (George Smith) Works. Selections. | Military art and science—United States. | Strategy. | Generals—United States—Biography. | United States. Army—Officers—Biography.
Classification: LCC E745.P3 D35 2016 | DDC 355.4—dc23 LC record available at https://lccn.loc.gov/2016021668

♾ Print editions meet the requirements of ANSI/NISO z39.48–1992 (Permanence of Paper).
Printed in the United States of America.

24 23 22 21 20 19 18 17 16   9 8 7 6 5 4 3 2 1
First printing

To Christina, my raison d'être

CONTENTS

ACKNOWLEDGMENTS

The greatest blessings in my life have been friends and family. Every success I have ever enjoyed has been a product of many hands. This book is no different. Brett Friedman first encouraged me to consider contributing to the 21st Century Foundation series. B. J. Armstrong is a kindred spirit, fellow history geek, clear thinker, and patient and thorough wordsmith. Glenn Griffith and the editorial and production team at the Naval Institute Press have improved my work and made navigating the publishing process a true pleasure. Bob McKenzie remains my moral, business, and legal adviser. My dad, John F. Daniel Jr., first exposed me to George Patton, forged my love for all things World War II, and taught me to value honesty and family. My mom, Mary Berry, still inspires my creativity, enables my book-hoarding-habit, reads drafts of my papers, and teaches me to "remember where you came from." Most important, I want to thank my best friend and wife, Dr. Christina Capacci-Daniel. To put it mildly, you are truly my raison d'être.

A Warrior's Honeymoon(s)

After marrying his childhood sweetheart, Beatrice Ayer, on 26 May 1910, George S. Patton Jr. and his bride sailed to England for their honeymoon. For Patton, who had grown up eagerly reading European history, this was his first trip abroad and far more than just a romantic vacation. After crossing the Atlantic in the bridal suite of the SS *Deutschland*, the Pattons landed in Plymouth and spent several days exploring Cornwall and the English countryside before traveling to London. Immersing himself in the history and traditions of Old Europe, the trip was a dream come true for George, who finally visited the castles and churches that had so captivated him in books.

While in London, Patton also made a series of unusual purchases for a honeymoon. He bought a number of rare books on military history and theory, including an early English translation of Carl von Clausewitz's *On War*. For Patton, this was the start of what would become a massive private library of books on military topics—and a hidden but essential part of his success.[1]

Although the newlyweds wanted to travel through Europe for several months, George was called back to Fort Sheridan, outside of Chicago, to resume his career as a young cavalry officer. A few weeks later, while on field maneuvers in Wisconsin, Patton wrote Beatrice about his reading, claiming that "Clausewitz is about as hard reading as any thing [*sic*] can well be and is as full of notes and equal abstruseness as a dog is of fleas."[2] Despite the hardscrabble realities of life as a junior officer, Patton often reflected on the palaces, villages, and countryside he had finally encountered—and the tangible connection to history he had felt—on his honeymoon.

The Pattons returned to Europe in 1912, when George competed in the modern pentathlon in the Stockholm Olympic Games. Despite his insistence on using his service pistol for the target shooting portion of the event rather than one of the specialized, smaller-caliber, target models his competition preferred, George still finished an impressive fifth overall. After the Olympics, and again in 1913, Patton was permitted to travel to the École de Cavalerie (Cavalry School) in Saumur, France, to take fencing lessons from the legendary swordsman Charles Cléry. Although he was very serious in his study of swordsmanship, Patton also considered this an opportunity to complete his honeymoon, and he traveled for several weeks in Germany and France with Beatrice.

While mastering the art of the sword in Saumur, Patton conducted a series of studies that became eerily relevant thirty years later. Reflecting on the military history of the region, most notably the campaigns of Henry I during the twelfth century, Patton reached the conclusion that the seemingly remote and insignificant territory in the Pays de la Loire region would play a major role in a future conflict. On his own initiative, he conducted a detailed reconnaissance of the local road network, and upon his return to duty at Fort Riley in 1913, he wrote a detailed report of his findings. This report, though accepted by the Army, would go unused for three decades until Patton himself relied on the report's roadmaps to help defeat the Nazis.[3]

Patton's honeymoons were in many ways a microcosm of his career and personal approach to the outside world. He was passionate about life and eager to experience it fully. More than simply a meditative observer, Patton was proactive in learning from the world around him, which ultimately prepared him for his almost inexplicable destiny. It is possible to infer too much from these early incidents, but they provide interesting insights into Patton's exceptional mind, character, and motivation as well as his ability to compartmentalize and retrieve critical knowledge at the right time.

The Challenge of Military Leadership

Modern military officers are presented with an incalculable number of personal and professional challenges. Successful leaders must possess a unique combination of character, physical strength, mental acuity, morality, interpersonal skills, confidence, and commitment to their profession. No other vocation in the world combines quite the same stress, uncertainty, and high stakes as a military career. To be a master of the military arts, one must be concurrently a doer and a thinker.

Today, after a decade and a half of near constant fighting beginning at the turn of the twenty-first century, there is a broad-based fear that the U.S. military faces a crisis in leadership and strategic thought. Some analysts critique the military personnel system, arguing that we do not recruit, promote, and retain the best human capital. Others are critical of military education and training, claiming that we do not adequately develop the mental and physical abilities of our leaders. Still others point to the difficulty of formulating effective strategy in a complex and uncertain geopolitical environment and fault military and civilian leaders for their lack of strategic insight. Whatever the specifics of the critique, at its foundation is a general sense of anxiety regarding the future of American military leadership.

One constant across all of military history is the critical role of people. Far more than technology, platforms, or luck, human assets have always been the single biggest determinant of military success or failure. As such, attracting, developing, retaining, and empowering skilled personnel is the biggest challenge for the modern military. One way to understand the issue is to look at exceptional leaders from the past and endeavor to learn from their experiences and choices.

Patton as a Man of Action

In almost every way, George S. Patton Jr. was exceptional. Despite his complex, multifaceted, and in certain aspects deeply flawed personality, Patton is correctly remembered as one of the finest leaders and military minds our nation has ever produced. Although Patton died seventy years ago, he is a man who deserves continued study.

Because of his improbable accomplishments and larger-than-life persona, Patton is a difficult man to appreciate. For better or worse, his image is now inseparable from the critically acclaimed 1970 film *Patton* starring George C. Scott.[4] Although the film, in a somewhat caricatured fashion, portrayed Patton as an anachronistic soldier seduced by the romance of war and tormented by an internal struggle to fulfill his destiny, there is a deep element of truth to this portrayal. Patton *was* a brash, profane, and energetic leader. Patton *was* a man convinced of the importance of his own destiny. Patton *was* one of the most daring, creative, and successful commanders of the twentieth century. However, because of Patton's stature as a great leader, it is essential to move beyond the Hollywood version of the man and focus on how he translated his desire for glory into success.

Physically, Patton was impressive. As a cadet, he set school records on the West Point track team and was a reckless and often injured member of the football team. While competing in the modern pentathlon at the Olympics in Stockholm, he earned the admiration of his fellow athletes for his fiery competitiveness. He was a world-class fencer who studied at the French fencing and cavalry school at Saumur, and the U.S. Army named him the first "Master of the Sword." Patton was also an excellent horseman and was generally regarded as one of the best polo players in the Army.[5]

He used his physical gifts and unnaturally high energy as part of his legendary leadership style. Seemingly never tired, he was always engaged and moving—a tempo that kept friend and foe alike guessing as to his next move and allowed him to dictate the direction and pace of operations. Patton led from the front and shared the same risks and hardships as his men. Despite a lifelong fear of being a coward, Patton was able to personally identify with and inspire his troops. His leadership and energy were infectious and helped motivate his men to achieve remarkable feats. Still today, veterans who served under him are proud to say, "I fought with Patton!"

As an officer, he augmented his energy and enthusiasm with a carefully crafted warrior image. Patton played the part of a soldier. He maintained the highest standard in dress and personal appearance, often changing uniforms multiple times per day in order to appear fresh. Committed to military standards, he nevertheless added custom elements to his uniform, such as his own ivory-handled pistols, a lacquered helmet, and fancy riding boots. Similarly, he practiced his "war face," salute, and posture in the mirror for hours and took personal delight in acting like a member of the officer class. Many thought this demeanor pompous or "too damn military," but it was an effective means of instilling esprit de corps in his troops.[6]

Patton's insistence on personal appearance and military deportment filtered down to his men, as he insisted on saluting—and wearing neckties and helmets—at all times. Likewise, he was quick to levy fines for improper military courtesy or uniforms. These high standards had an immediate effect on his men, who commonly referred to a snappy salute as "a George Patton."[7] Although some saw this fixation on appearance as unnecessary in modern war, Patton justified it as an essential part of a soldier's training and critical for forging unit cohesion.

Throughout his career, Patton was an enthusiastic proponent of training both the individual and the unit. He insisted his men meet rigorous physical standards and attain expert proficiency in the use of weapons and the specialized technical skills associated with modern warfare. He demanded much from his troops, but he was careful to support them, ensuring that they had the best quality food and clothing available and that their equipment was up to authorized strength and properly maintained. In addition to this individual training and support, Patton built unit-level skills through both parade drills and war-game maneuvers. To ensure that the lessons of the field exercises were absorbed, he frequently gave special lectures and debriefings in his characteristically flamboyant, impassioned style.

Dedication to training was a key element in Patton's career. He achieved impressive results in a series of 1941 field exercises in which his troops ran circles around their opponents, thus demonstrating the need to master fire and movement in the coming war. After the American entry into World War II in December 1941, Patton's first assignment was to create the Desert Training Center, intended to rapidly instruct troops in the rudiments of desert combat and survival. Nothing could completely prepare these soldiers for the battles that would come in North Africa, but Patton did his best to instill urgency and purpose in their training. As he often said, "A pint of sweat will save a gallon of blood."[8]

In combat, the results of Patton's leadership were clear. During the Punitive Expedition in Mexico, World War I, and World War II, he excelled across a wide range of contingencies and levels of command.

When opportunities presented themselves, he acted decisively and aggressively, believing that his destiny was to achieve military greatness or die trying. Patton actually suffered from severe self-doubt regarding his leadership and courage, but he forced himself to move forward and act boldly, saying, "Do not take council of your fears" and quoting Fredrick the Great's maxim that one should always act with audacity. This hell-for-leather dynamism inspired others to follow him and was a true force multiplier. Patton recognized that *"leadership* is the thing that wins battles. I have it, but I'll be *damned if I can define it."*[9]

As a man of action, Patton was successful because he was able to combine a unique set of physical gifts, a dynamic personality, a carefully crafted warrior image, commitment to the training and personal welfare of his men, and the intangible qualities of leadership. But this passion for action, though critical for Patton's success, also contributed to his failures. As a commanding officer described Patton in a fitness report, he was "invaluable in war . . . but a disturbing element in time of peace."[10]

Indeed, Patton struggled for his entire career to find the correct balance between aggressiveness and discretion. His aggressive spirit and thirst for action at times impeded his judgment and resulted in many of his troubles, including a series of poor officer evaluation reports in the 1920s; his rocky relationship with his commanding officer in Hawaii, Gen. Hugh Drum; the slapping incidents in Sicily during World War II; and his failure to adapt to the realities of the postwar world as pro-council to Bavaria. Had Patton served in a later era, it is likely that his career would not have survived any one of these incidents. Even in his own time, Patton was lucky to have had friends who recognized his talents and protected his reputation and career. His fire and drive, though critical elements of his success, must be understood as nearly tragic flaws.

The Patton Mythos

One of the most difficult challenges in studying Gen. George Patton is separating the myth from the man. Popular depictions of Patton typically focus on his larger-than-life persona as a man of action. According to these depictions, Patton was misunderstood by jealous peers who could not measure up to his greatness, acted impulsively without thought or regard to protocol or politics, was an anachronistic and romantic warrior trapped in the past, and had an innate and mystic ability to instinctively comprehend military problems. Although there is an element of veracity to each of these claims, they form, at best, an incomplete picture that does not capture the whole truth about Patton.

In truth, Patton had excellent personal relationships in military and civilian circles. He cultivated close and valuable friendships and mentors throughout his career, including powerful and influential figures such as John Pershing, Henry Stimson, George Marshall, Dwight Eisenhower, and Omar Bradley. During multiple assignments at Fort Myer, Virginia, Patton used his combination of charm and family money to adroitly navigate the social and political milieu of the interwar Army. There, he developed lasting friendships with Stimson and Marshall, and those friendships paid tangible dividends, as they recognized Patton's talents and promoted and protected him throughout his career. Although Patton may have preferred the fields of battle to polite society or the halls of power, he was an atypically adept socialite and politician. He undermined his career with a series of gaffes and lapses in judgment, such as the Knutsford Incident in England and various public statements during his tenure as military pro-council to Bavaria, but otherwise, he typically acted in a restrained and careful manner. In fact, the very incidents that created Patton's image as an impulsive, political buffoon were the exceptions to the careful political calculation and maneuvering that defined the vast majority of his career.[11]

Another one of the most persistent Patton myths was that he was a man who was trapped in a romantic vision of the past from which he could not escape. Patton is at least partially responsible for this myth, as he cultivated the image of an old-fashioned cavalryman and frequently made obscure historical references. But despite his atavistic persona, Patton was thoroughly modern. With his lifelong love of speed and technology, he was an early adopter of the radio, aircraft, and automatic weapons in warfare. During the 1916 campaign in Mexico, he led the first mechanized assault in American history when he raided a house occupied by one of Pancho Villa's top commanders, Julio Cárdenas.

His rapid adoption of the latest technologies continued during World War I, when he transferred to the Army's armored branch. In his new position, Patton learned all he could about tanks and mechanized equipment. He established a school for tank crewman and led one of America's first armored units into battle. Continuing this work during the early interwar period, Patton consulted with American tank designer Walter Christie to solve some of the technical problems he observed through firsthand experience in France. Also during the interwar period, he earned a pilot's license because he was convinced that light aircraft would provide a more accurate and timely view of the battle space and would allow him to shuttle rapidly between key sectors of the front. Further, Patton eagerly embraced modern infantry weapons, particularly the M1 Garand. He believed that the new rapid-firing rifle would provide the firepower necessary for his forces to regain the ability to advance across open ground, thus restoring movement to the battlefield. In these and countless other technical matters, Patton's interests were both practical and theoretical. He enjoyed working on engines and machinery as well as thinking about how these new technologies affected the conduct of armed conflicts. He had a deep love of history and tradition, but he was much more forward looking than popular legend suggests.

Finally, Patton is commonly depicted as an instinctive man of action with an innate sense of purpose and an almost mystical connection to the past. According to this view, he was a natural warrior because he had an instinct for battle that provided him a unique ability to pierce the fog of war and see clearly and effortlessly where others could not.[12] Although Patton did have a natural gift for military leadership, he in fact worked diligently to cultivate his seemingly instinctive talents.

From his earliest days, Patton had a passion for poetry and military stories. As a boy, family members would read to him from military classics and regale him with stories of glorious ancestors and great captains of the past. One of Patton's favorite storytellers was family friend and Confederate partisan leader Col. John Singleton Mosby. The old veteran would keep the young boy enthralled with stories from the Civil War and impart lessons on horsemanship and conduct under fire that Patton would repeat and employ decades later. During his shootout with Cárdenas and the Mexican banditos in 1916, Patton recalled that Mosby always said to first shoot at the horse of a mounted fighter, not the man himself. This advice proved immensely valuable. In an attempt to stop one of Cárdenas' men from fleeing the scene of a shootout, Patton recalled this advice and was able to first stop the man's horse and then eliminate the bandito.[13]

In spite of a deep desire to learn more about military subjects, Patton was actually a poor student in his youth. He was slow to learn how to read and write, so much so that many later biographers would speculate that he suffered from dyslexia. Despite his academic struggles, Patton worked diligently his entire life to improve upon and develop his innate military acumen. By force of will, he became a dedicated student and overcame his initial scholastic challenges. He concentrated his efforts on reading, focusing primarily on subjects he enjoyed, such as poetry, history, and military theory.

Patton was fortunate to have received an impressive formal military education from the Virginia Military Institute, West Point, the École de Cavalerie, the U.S. Army Command and General Staff College, and the U.S. Army War College. Despite this impressive list of academic credentials, Patton's military knowledge was primarily the product of his own efforts. Like many officers of his era, he believed that Army schools taught an outdated and inflexible curriculum that forced students to adopt textbook solutions to problems, a process that stifled creativity and originality.[14] Patton did well academically at these institutions, but he was convinced he needed to conduct his own studies to fully grasp complex military problems.

On his own initiative, Patton studied works of military history and theory because he believed that they contained valuable truths about the nature of war. He read, reread, meditated on, and made extensive notes detailing his observations, using these notes over time to study, internalize, and refresh his memory on key points. Because of his dedication to reading and study, Patton amassed an impressive personal library of military books, which included the early English edition of Clausewitz's *On War* that he purchased on his honeymoon. He frequently consulted his library and took great pains to move it with him from post to post during his career.[15]

Patton believed he was destined for military greatness but recognized that academic knowledge would be critical to fulfilling his potential. As a result, he formed himself into a deep and original thinker about a wide range of theoretical and practical military problems. This commitment to self-improvement distinguished him from the majority of his peers, who failed to dedicate themselves to such a rigorous regime of study and reflection. Ultimately, Patton was so advanced in the military arts that it seemed as if he had a natural and instinctive knowledge of the subject, an appearance that belied his purposeful attempt to master his chosen profession.

Patton as a Thinker?

Contrary to his reputation as an instinctive man of action, Patton was a dedicated and lifelong student of military history and theory. His studies led him to consider both enduring philosophical questions about the meaning of war and more practical issues regarding the roles of new technologies and the future of warfare. In his spare time, Patton composed detailed private notes, poems, diary entries, letters to friends and colleagues, professional journal articles, lectures for his men, official reports, and a thesis for his graduate work at the Command and General Staff College and War College. The breadth and depth of thought in these writings is truly impressive.

During his career, Patton studied and wrote about many of the major military questions of his day in an attempt to better prepare himself for his martial destiny. He recognized that advances in mechanized and combined-arms technology were changing the pace and ferocity of warfare and concluded that interwar Army doctrine was an outdated product of a bygone era. Weapons such as tanks and airplanes were providing increased opportunities for breakthrough and pursuit, invalidating many commonly held assumptions about the value of protecting flanks and the dominance of the defense.

These changes would require a fundamental revision of strategic thinking and tactical doctrine. To account for this revolution in military affairs, Patton also saw the need for military and bureaucratic reorganization. Such restructuring necessitated rethinking the traditional role of cavalry, creating an independent tank corps, and a closer cooperation between frontline troops and their air and artillery assets. Patton was a product of the horse cavalry, but he came to believe that the interwar Army's organization was anachronistic and in need of wholesale reform.

Much of Patton's literary efforts during the interwar period attempted to convince the Army that the nature of war was changing and the service was in danger of becoming obsolete. Patton believed that industrialization had deepened the impact of warfare on society, correctly anticipating the total nature of World War II. Similarly, he predicted the resurgence of Germany and studied with keen interest the advances in combined arms and Blitzkrieg-style warfare. Finally, in perhaps his most prescient analysis of the future, he conducted a series of studies about the potential for a Japanese attack on the Hawaiian Islands. In 1937 he authored an intelligence report that predicted a surprise attack on U.S. forces in Hawaii using carrier-based aircraft. It proved to be an eerie harbinger of the Pearl Harbor attacks.[16] Although many of Patton's predictions during the period appeared shrill and alarmist at the time, in hindsight they were remarkably accurate.

Because of Patton's achievements on the field of battle, his genius as a military thinker is often overlooked, an unfortunate oversight that fuels the stereotypical views of the general that persist to the present day. An intense man of action, he was an equally intense man of thought. A more complete appreciation of Patton leads us to better understand the intellectual groundwork for his future successes and to appreciate him more fully—and accurately—as a true military genius.

Patton's Relevance Today

Providing a more nuanced portrait of Patton is interesting, but it is not the primary goal of this work. Rather, this book hopes to use the lesser known, intellectual side of Patton's character to highlight his contributions to the enduring debates involving military affairs and strategy. Contemporary strategists often have difficulty translating the insights of past thinkers into something that is applicable to current debates. Fortunately, Patton left a rich body of written work that boldly presents his vision of warfare and provides perspective on modern issues.

Although Patton's writings were focused on solving the tactical and strategic problems of the early to mid-twentieth century, several other themes clearly emerge from his texts: (1) the importance of leadership and the warrior mindset to achieving military success; (2) the role historical, cultural, political, and technical knowledge plays in the military profession; (3) the need for education and training to develop effective leadership; (4) the tension between the ancient principles of warfare and the impact of technology on the conduct of war; and (5) the importance of critical thought about complex problems in a competitive strategic environment.

During his career, Patton struggled with each of these complex topics in a purposeful and intellectually rigorous manner. Rather than just an exercise in self-improvement, Patton believed that this search for understanding was essential if he was to fulfill his destiny as a great military leader. Ultimately, he was successful because he was able to express his vision of leadership in a manner that enabled others to learn from and be inspired by his example. Patton's military writings provide modern audiences a glimpse into this lesser-known side of Patton's character and allow for his strategic insights to endure.

The remainder of this book demonstrates Patton's continued relevance by providing some of his best writings on the military and strategic arts. Each chapter begins with a brief introduction that traces the intellectual and professional development of Patton while placing the writing within its proper historical context. The remainder of each chapter is the complete and unedited words of George Patton.

Chapter 1, "An Early View of a Military Mind," presents Patton's March 1913 *Cavalry Journal* article, "The Form and Use of the Saber." This chapter explores Patton's earliest attempt to publish his theories on war. Although Patton's thoughts were still in their infancy at the time, the article provides fascinating insights into the intellectual development of the future general as well as some guidance on swordsmanship. In this brief article Patton attempts to combine his training in swordsmanship at the French cavalry school with his already impressive knowledge of military history. Despite the anachronistic subject matter, in the tradition of Miyamoto Musashi's classic text of samurai swordsmanship, *The Book of Five Rings*, this article offers a synthesis of practical knowledge and experience with theory and historical examples that offers lessons beyond edged weaponry.

The focus of chapter 2, "The Warrior Mindset," is on Patton's 1927 lecture titled "Why Men Fight." Here, Patton attempts to apply his experiences in World War I to grapple with difficult questions regarding leadership, loyalty, and courage. For Patton, these qualities were constant across time and the biggest determinants of success or failure. He had written privately on this topic for two decades prior to the lecture, but this bold statement of the principles of leadership reflect his growing belief in his own destiny.

Chapter 3, "Technology and War," introduces Patton's article from the November 1930 issue of *Cavalry Journal*, "The Effect of Weapons on War." Despite his love of history and old-fashioned weapons such as the sword and horse cavalry, Patton was an early adopter of technologies such as automobiles, airplanes, tanks, and modern infantry weapons. In this article, Patton expresses the tension between the past and the need to adopt new technologies while retaining a proper sense of history and tradition.

Chapter 4, "Patterns of Success," is based on Patton's article, "Success in War," which appeared in the *Cavalry Journal* in January 1931. Throughout his career, Patton searched for common themes in military history that led to victory. Here, Patton elucidates his vision for replicating success in future wars in his characteristically bold style, making excellent use of historical analogies to distill those patterns.

Patton's 1932 War College thesis, "The Probable Characteristics of the Next War and the Organization, Tactics, and Equipment Necessary to Meet Them," forms the basis of chapter 5, "Anticipating the Next War." This thesis is remarkable because it shows Patton's ability to apply his knowledge of history and military affairs in order to anticipate the key elements of future conflicts. Indeed, much of Patton's legendary ability to appear at the right place at right moment can be traced to this exceptional intellectual talent and strategic foresight.

Expanding on this ability to anticipate the characteristics of future conflicts, chapter 6, "Refining the Concept of Mechanized War," presents a 1933 article from the September–October issue of *Cavalry Journal*, "Mechanized Forces: A Lecture." Here Patton focuses his attention on the development of armored warfare. Although he was one of the key organizers of American armored forces in World War I, he rejoined the horse cavalry during the interwar period because of the lack of funding and opportunity for advancement. Despite this pragmatic career move, Patton was a true believer in the role of an independent armored force in future conflicts, and he kept current on developments in tank technology and tactics. He believed that armored warfare would play a decisive role in future combat, and he passionately advocated for an independent and modern tank corps. Patton never received the credit of other armored theorists from the period, but this oversight should not diminish his accomplishments as an armored visionary.

Chapter 7, "Training the Force," provides a final view of Patton's evolving study of war, "Desert Training Corps," from the September–October 1942 issue of *Cavalry Journal*. In this short work, Patton provides a discourse on the value of realistic combat training and highlights his role in establishing the Desert Training Center. Though he had yet to lead his men into battle during World War II, his confidence and clarity of purpose are striking. Despite its short length, this article is a fitting final work for this book because it is a culmination of Patton's lifelong quest to distill the lessons of war into a useable set of beliefs and values.

The book's conclusion introduces "A Soldier's Reading," a short essay by Patton's wife, Beatrice, that appeared in the November–December 1952 issue of *Armor*. As Patton's most trusted friend and intellectual sparring partner, Beatrice had unique insight into Patton's professional development and reading habits.

Finally, to assist readers who may wish to study General Patton in greater detail, the "Further Reading" section provides a bibliography of the most useful Patton sources.

ONE

An Early View of a Military Mind

Contemporary military leaders are faced with a dizzying array of technical and theoretical information. The U.S. military does its best to provide training and mentorship in the field and in the classroom, yet it is often difficult for even the most diligent warrior to properly reflect on lessons learned and apply the insights thus gained to future contingencies. Although globalization and the increased speed of information dissemination may have altered the specifics of the problem, the broader issue is not new.

Part of Patton's genius was his ability to think critically on many different levels about military challenges and to distill his insights into a coherent set of lessons. Even as a junior officer he was keenly aware of the need to develop both practical and theoretical knowledge, and he worked throughout his career to expand and deepen his knowledge of weapons, tactics, military history, and strategy. Patton's dedication to being a complete soldier resulted in a purposeful process of self-improvement. His studies allowed him to seamlessly transition between tactical action and strategic thought and contributed to the mythos of Patton as a natural warrior. In fact, much of Patton's success was the result of his unique ability to study and synthesize military knowledge into a usable set of principles.

Although from an early age Patton privately recorded his thoughts on military subjects, his first published work was

"The Form and Use of the Saber," which appeared in *Cavalry Journal* in March 1913. Here, Patton provides a first glimpse at his impressive military acumen. Based on his recent study at the École de Cavalerie at Saumur and his deep knowledge of military history, Patton critically examined the U.S. Army's history, training, and implementation of the cavalry saber and came to a startling conclusion: the Army had the wrong weapon and doctrine.

In the conservative pre–World War I Army, such a challenge to tradition was a significant risk for the young lieutenant. To counteract the deep loyalty within the Army for the traditional saber model, Patton needed to carefully support his reform efforts with both technical expertise and a knowledge of history. In both of these tasks, Patton was uniquely suited to the challenge. Through his synthesis of practical knowledge, military history, and theory, he made a convincing case that a new saber design and doctrine were long overdue. Patton began his efforts with a skillful lobbying of key members in the Army bureaucracy, most notably Gen. Leonard Wood. He also cultivated contacts in the Army Ordnance Department, where he demonstrated the value of his proposed changes with live demonstrations of his proposed changes.[1]

This March 1913 article is a written summary of his lobbying efforts, which were designed to transmit his ideas to a broader audience. In it, Patton argues that the Army must have the best equipment and doctrine possible. He then explains the two theories regarding swordsmanship—the use of the point versus the use of the edge—and demonstrates the relative advantages and disadvantages of both methods. Next, he summarizes the previous millennia of military history and concludes that striking with the point of a weapon is generally preferable to slashing with the weapon's edge. Because hitting with the point of the blade is preferable, Patton then asserts that the U.S. saber, designed to strike with the curved edge, is outdated and in need of replacement. Finally, Patton concludes his argument with a brief set of suggestions on how to best use the proposed straight-bladed saber in combat.

His efforts paid off surprisingly quickly by today's standards, as the Army soon adopted the Model 1913 cavalry saber, which became affectionately known as the "Patton sword." This new weapon featured a straight blade optimized for stabbing with the point rather than a curved edge for slashing, just as Patton had suggested. Patton was briefly detailed to the Springfield Armory in Massachusetts to inspect the initial production run of these new blades and assure that they met his demanding standards.

Patton was satisfied with the quality of the new blades, but he still had to ensure proper implementation of the redesigned weapons. In the summer of 1913, he traveled, at his own expense, back to Saumur to take further instruction in the use of swords. He returned to the United States in the fall and was assigned to Fort Riley, Kansas, and given the title "Master of the Sword." In this new role, Patton instructed officers in the proper techniques for wielding the new sword and consistently impressed them with his knowledge and passion. Although most of the officers he taught were senior to him in rank, he was careful to parry potential jealousies through an impressive combination of tact, humor, and a true mastery of the subject. To ease the tension, Patton would begin the first class with his new pupils by showing them a wooden sword he had used since childhood, joking that he had been studying the art of swordsmanship since he was a boy, and noting that only in the field of swordplay was he senior to them.[2] The final contribution Patton made to swordsmanship in the U.S. Army was the creation of a new manual of arms for his redesigned saber.[3] This document, published in 1914 under the title *Saber Exercise*, formalized and expanded upon the points he made in his 1913 article.

Although the focus on edged weapons may appear quaint in retrospect, the change to the new sword was a significant accomplishment for First Lieutenant Patton and provides a fascinating glimpse into Patton's character and mind. Patton was able to achieve this rapid success because he took risks and acted proactively, made key political connections, lobbied skillfully for his proposal, and was able to support his ideas with a unique combination of practical and academic knowledge. This general pattern set the course for Patton's later successes. As such, Patton's first attempt to formalize his thoughts for a public audience is worth continued study and consideration.

THE FORM AND USE OF THE SABER

Cavalry Journal, March 1913

At first sight it seems rather curious that, though the saber has been a component part of our cavalry equipment ever since the beginning, its use and form has never been given much thoughtful consideration. When we consider, however, that for years the only target practice our troops had was when the old guard fired the loads from their muskets, our negligence in acquiring other knowledge seems less strange. It was through the personal interest and excellence of individual officers and men that attention to target practice was first introduced. I have been informed by some of these gentlemen that at first they were met by obstructionists and the cry of "let well enough alone." They persisted, however, and as people began to see the results they accomplished they ceased to hinder, and rapid and wonderful progress both in the rifle and in the manner of its use have followed.

It now seems that the turn of the bayonet and saber has arrived. But to gain any prominence it must be supported by some personal interest on the part of officers and men which, when applied to the rifle, has given us the greatest shooting arm in the world. Yet, however essential this interest may be it is difficult to excite it with our present saber and methods of instruction.

As to the form of the saber, there seems to have been an age-long controversy between the advocates of the edge and those favoring the point. Beginning with the 11th Century, from which time accounts are fairly consecutive, we find as follows:

When scale, and later chain, armor became sufficiently perfected to completely cover the body, the point went out of use because it was quite impossible to thrust it through the meshes, while by giving a violent blow, it was possible to break or cripple an opponent's arms or ribs without cutting the armor.

When the German Mercenaries in the Italian wars began to wear plate, the Italians found the edge of no avail and returned to the point which they thrust through the joints of the crude plate armor. Gradually armor became so well made that neither the point nor edge affected it, but about this time the bullet began to put the armor out of business.

While the armor was being eliminated, so-called light cavalry was evolved. These men wore no armor, and since the Cossacks, Poles, and Turkish horsemen were the only examples of the unarmored horse that men had to copy, and because these inherited from the Arab a curved, scimitar-like saber, the new light cavalry was mostly armed with a curved saber. The weapon adopted was, however, an unintelligent copy. The scimitar of the Oriental was a special adapted for cutting through defensive clothing made of wool wadding and to be used in combats when the opposing horsemen fought in open formations circling each other and not in ordered lines trusting to shock.

The sword given to most of the light cavalry troops was not of sufficient curvature to give the drawn saw-like cut of the scimitar and yet was curved sufficiently to reduce its efficiency for pointing. It may also be noted that the scimitar was not used for parrying and could not be, having neither guard nor balance. All the parrying was done with a light shield. But this lack of balance and the curved form of the weapon must not be considered as essential to a cutting weapon, for the long, straight, cross-handled sword of the Crusader has a most excellent balance, about two inches from the guard. Yet this weapon was probably the one of all time capable of striking the hardest blow.

The present saber of our cavalry is almost the last survival of the incorrect application of the mechanics of the scimitar. It is not a good cutting weapon, being difficult to move rapidly. It is not a good pointing weapon, being curved sufficiently to throw the point out of line. Yet it is clung to as fondly as was the inaccurate Civil War musket and the .45 Springfield with its mule-like kick.

The tenacity evinced for the retention of an illogical weapon seems without basis in history, while from the same source we find numerous tributes to the value of the point. Verdi du Vernois says, "Experience has shown that a sword cut seldom, but a point with the sword always, throws a man off his horse."

In the Peninsula War the English nearly always used the sword for cutting. The French dragoons, on the contrary, used only the point which, with their long straight swords caused almost always a fatal wound. This made the English say that the French did not fight fair. Marshal Saxe wished to arm the French cavalry with a blade of a triangular cross section so as to make the use of the point obligatory.

At Wagram, when the cavalry of the guard passed in review before a charge, Napoleon called to them, "Don't cut! The point! The point!"

To refute this and much more historical approval of the point and the present practice of all great nations, except Russia, the advocates of the so called cutting weapon say that we are practically a nation of axmen. It is doubtful, however, if many of our men have ever handled an ax or are descendants from those who have. The tendency of the untrained man to flourish his sword and make movements with it simulating cuts is to be found in other nations. In France, noted for its use of the point, I witnessed within the last year several hundred recruits, when first handed sabers, thrashing about with them as if they were clubs, but no sooner were they taught the value of the point than they adopted it and never thereafter returned to the edge.

The child starts locomotion by crawling, but on this account do we discourage walking? The recruit flinches and blinks on first firing a gun, but he is certainly not encouraged to continue this practice. Why, then, should the ignorant swinging about of a sword be indicative of its proper use? It is in the charge that the sword is particularly needful, and, in fact, finds almost its whole application, and it is here that the point is of particular advantage in stimulating to the highest degree the desire of closing with the enemy and running him through.

In executing the charge with the point, according to the French method, the trooper leans well down on the horse's neck with the saber and arm fully extended and the back of the hand turned slightly to the left so as to get the utmost reach. This also turns the guard up and thus protects the hand, arm, and head from thrusts and the hand from cuts. The blade is about the height of the horse's ears, the trooper leaning well down and in the ideal position slightly to the left of the horse's neck. In this position he can turn hostile points to the right by revolving his hand in that direction, the point of his weapon still remaining in line and he himself covered by the guard of his saber. The pommel of the saddle and the pommel pack, such as is on our new saddle, protects the thighs and stomach from points deflected downward. Cuts would fall on the shoulder or across the back where they would be hindered by clothing and to little harm. The head can be protected by ducking it below the horse's crest. Moreover, since the point will reach its mark several feet before a cut could be started, there is little danger of its being dealt. Should it be necessary to attack an opponent on the left, the arm is brought over the horse's neck and the hand rotated further to the left, keeping the guard before the face. In this position the parry for the point is either up or to the left.

Another advantage of this position is that while pushing forward to close, only half the human target visible in our present position of charge is exposed, and that in urging the horse to speed the best results are attained with the weight carried forward as described. To use the edge it is necessary to sit erect and in the act of dealing a cut the trooper is completely open either to cuts or thrusts. Moreover, his reach is shortened at least three feet, for the cut to be effective must be dealt with the "forte" of the blade which starts about eight inches from the point and in a position to cut the trooper also loses the entire reach of his extended body and arms.

The point is vastly more deadly than the edge, for while it might be possible to inflict a crippling blow with the edge were the swing unrestricted by the pressing ranks of the charge or by the guard or attack of an adversary, yet with both of these factors added to the necessity of so starting the cuts as to reach its mark after making due allowance for the relative speed of approach of the two contestants; the size and power of the blow becomes so reduced that there is grave doubt if it would have sufficient power to do any damage to an opponent's body, protected as it is by clothing and equipment. And even should it reach the fact, its power to unhorse is dubious.

The use of the point, on the other hand, is not restricted by the press of the ranks and its insinuating effect is not hindered by clothing or equipment. The exaggerated idea of the effect of a cut which prevalent in our service is due possibly to the fact that when a man wants to demonstrate it he rides or walks up to a post, and with plenty of time to estimate distance and with his swing unimpeded by companions on either hand, he can expend all of his power and attention to chopping at his mark. Also, in our so called fencing, mounted or dismounted between enlisted men, the touch with the point which, were it sharp, would introduce several inches of steel into its target, is hardly felt, while blows with the edge often cause considerable bruises, though were these sharp it is doubtful if they would do more.

It is also well to remember that were one of our lines, charging as at present, to run up against a line charging with the point, our opponents' weapons would reach us and have ample opportunity to pass through us before we could be even able to start a cut in return. Were we, on the other hand, while using the point, to encounter men using the edge, we in turn would have them at our mercy. In the melee which follows a charge, there is less objection to using the edge, for the horses will be going at less speed and things will probably open up. At least, there will be no rank formation and a man can chop away as ineffectually as he likes, though here, too, the point would be more deadly. In the pursuit there is little choice between the edge and point, though it might be a little easier on the horses to stick a man when he is several feet ahead than to be forced to ride almost abreast of him to deal a cut. Moreover, a man can parry a cut from behind while continuing his flight, but in order to parry a thrust he must stop and turn. Still, with the straight sword under consideration by the War Department, cuts can be more effectually made than they could with our present saber, as the new sword is better balanced for rapid cutting and is very sharp on both edges. Of course, this weapon is distinctly a cavalry arm, and it would not affect the equipment of the infantry or artillery.

In instructing the trooper in the use of the saber, he is never allowed to fence with beginners but is assigned to a noncommissioned officer or an instructed private who teaches him the mechanism of the thrust and the idea of parrying with the blade while keeping the point in line and always replying to an attack with a thrust. Later, he is allowed to use occasional cuts, but he is always impressed with the idea of thrusting. This instruction will give him facility in the use of his weapon and impress him with an aggressive spirit. He is then placed on a wooden horse and first taught the position of charge, mounted, and how to parry with his blade while in the charging position without getting his point out of line for his opponent's body. He is then placed on horseback and taught to take the proper position and later to run at dummies of considerable weight. In running at dummies, there is no jabbing with the arm. The blade is kept still and the horse does the work. All the man has to do is to direct his point, which operation is facilitated by the fact of his having his blade along the line of sight. Later he is taught to use his weapon against adversaries on his right and left as in a melee. In teaching this he is first allowed to go slowly, but having learned the mechanism he is thereafter required to go fast and is never permitted to slow up or circle. He rides at a man to kill him, and if he misses, he goes on to another, moving in straight lines with the intent of running his opponent through.

As to the question of recovering his sword thrust into an opponent, it is not difficult with a dummy when the latter is given any flexibility at all, and when a man has been run through he is going to be pretty limp and will probably fall from his horse, clearing the weapon for you. It would seem, then, that the straight sword possesses all of the advantages of the curved sword for cutting, besides admitting of the proper use of the point, which the other does not, and that in using the point in the charge not a single advantage of the edge is lost, while many disadvantages are overcome. In addition, the highest possible incentive to close with the enemy is given.

Finally, many of our possible opponents are using the long straight sword and the point in the charge. To come against this with our present sabers and position of charge would be suicidal.

TWO

The Warrior Mindset

Recruiting, inspiring, and leading men and women to sacrifice for the greater good of their society has been an enduring challenge for nations. Just as the fortunes of ancient Rome rose and fell on the backs of its soldiers, modern societies rely on the courage and sacrifice of their citizens for survival. This need for sacrifice and leadership is particularly strong in modern democracies that rely on all-volunteer forces. The willingness of free citizens to forgo a measure of their freedom to protect the rights of others is a special thing and by no means guaranteed.[1] Once these service members join the ranks, there remains the question of keeping them motivated and prepared for combat. For many, the solution to these problems is to create a class of warriors within a society, but this solution seems contrary to the egalitarian impulses of democratic societies, creating the very real possibility of a civil-military divide between those who serve and those who do not. Understanding the motivations and dynamics of the warrior mindset is thus a critical task for soldiers and citizens alike.

George Patton was deeply concerned about these questions, particularly during the interwar period, when the U.S. Army was decimated in terms of budget and manpower. He understood the value of excellent soldiers from his experiences in World War I and saw the development and retention of motivated soldiers as a matter of national security. Because

Patton believed that the Army was in dire need of effective leadership, he dedicated a substantial amount of private study to these issues.

In October 1927, Patton expounded upon the subject of leadership in "Why Men Fight," a lecture that provides a fascinating insight into Patton's career and philosophy.[2] He delivered this lecture after he had been stationed in Schofield Barracks in Oahu, Hawaii, for about a year, serving as the post supervisor for operations and training. Patton was unhappy with this assignment and disturbed by the casual nature and pace of life on the island outpost. After voicing his concerns regarding the readiness and leadership of this unit, he was quickly reassigned to serve as a base intelligence and security officer. Here, Patton again ruffled feathers by pushing too hard and appearing like a firebrand and was reassigned to Washington, D.C., to serve in the Office of the Chief of Cavalry.[3]

Despite the potential damage to his career, Patton could not tolerate uninspired leadership and insisted on imparting his style to the units in which he served. This intolerance would reemerge in an even more extreme manifestation when Patton slapped two of his soldiers during the 1943 Sicily campaign. Later, when describing the slapping incidents, Patton stated that he truly believed he had served the best interests of the soldiers involved and the Seventh Army as he had acted to force cowards back into a fighting mindset. Although Patton has been harshly criticized for his actions in Sicily, those actions are strikingly consistent with views he expressed sixteen years earlier in this lecture.

In this article, Patton uses his knowledge of history and military theory to draw on a wide range of analogies, ranging from ancient Greece to World War I. He concludes that, throughout time, warriors have fought for the same basic reasons: "hunger; sex and its simpler derivatives; unity of action, due to unity of impulses; biologically produced leaders; greed; the need for protection; ambition; romance; monotony and habit." For Patton, the challenge for contemporary leaders is to use these enduring truths to harness and exploit these basic impulses.

Throughout his career, Patton practiced what he preached in this lecture. Believing that officers must expect high standards of physical, personal, and professional deportment from their men, he set and enforced a strict code of behavior. He took an active part in the punishment and reward process and was quick to levy fines for minor infractions of discipline, but he also carried a collection of medals to personally reward troops for exceptional service. Patton acted the part of a great captain of history, evoking the spirit of a warrior caste with his fancy deviations from uniform standards and weapons. To reassure his troops, Patton shared the same risks as they did, frequently visiting the front and enduring bombing and shellfire. Because he knew that soldiers fought for physical necessities, Patton made sure that his men were provided with excellent food, clothing, and medical care. Finally, Patton understood inertia and realized it is much easier to keep men moving once they are in motion rather than get them moving after they have stopped. This belief contributed to his desire to win battles through aggressiveness and daring and to make certain his troops never stopped moving forward.

Today, Patton's lecture appears both dated and prescient. While his views on race, class, and sex are an outdated reflection of his particular era, he adroitly highlights the ongoing need for military sacrifice and leadership. Similarly, he highlights many of the difficulties in creating a civil-military balance that serves the needs and desires of a democratic nation. Patton's observations about the value of drills, pomp, self-esteem, unit cohesion, education, leadership, and a sense of purpose are fundamentally sound. His ability to internalize the individual components of leadership and military spirit is part of Patton's genius and directly contributed to his wartime success. Although clearly imperfect, this essay provides a critical insight into Patton's understanding of this important subject.

WHY MEN FIGHT

1927

With the causes and effects of war we are not concerned. Its continued existence is inevitable and its results for good or evil are beyond all human power to avert or change.

Our effort here is rather to seek those fundamental emotions which actuate men as individuals to expose themselves to wounds and death; to trace the growth and development of these emotions and finally to investigate how best they may be utilized and stimulated so as to produce in our armies that fighting spirit which will spell victory in the wars which are to come.

Of prehistoric man we know little, and can but opine that in his infancy he was a fairly strong and relatively intelligent, carnivorous animal.

Speaking generally, present day carnivora are not cannibals. Their innocence in this respect, however, is not the result of qualms of conscience, but arises from the fact that aware of their own powers they respect those of their kindred. Yet, since they must eat to live and kill to eat, they destroy for choice beasts less mighty than themselves and only combat their kindred when driven by hunger, first, to wrest from them their prey, and second, in the stress of famine, to devour them for food.

In primitive man then it would, by analogy, seem all but certain that the primary emotion inciting to combat with his fellows was the instinct to survive—the belly lust.

The next most powerful emotion inducive to fratricidal strife was sex. In its simplest form this incentive has lost its potency so far as civilized soldiers are concerned, but it is beyond question that it still exercises a strong influence among backward peoples, as, for instance, Mexican bandits.

On the other hand, it is clear that some derivatives of the sex emotion still retain the chief place among the inducements to combat.

In tracing these derivations we must begin at the dawning.

The necessity of fighting for the acquisition and possession of his mate gradually awakened in the budding intelligence of man an enhanced notion as to her value. This in the course of ages limited promiscuous breeding and engendered ideas of a permanent family.

To defend his harem, man fought his fellows. While with the increase of permanent ties his roamings were limited; resulting eventually in the establishment of a cave or den home. Long usage developed the idea that his particular hearth was the best of its sort in the neighborhood while conditions of intense cold gave further point to the notion by the necessity they imposed of having permanent and warm sleeping quarters. To defend his females and his bed, man fought for his home. In other words, the lust of sex eventually evolved into a habit of thought founded on ownership and the obligation to defend it. So was the germ of patriotism conceived.

Eons rolled by and in their course reason and usage developed a tolerance for the young males until at last they were permitted to remain. Since each of these males had in him the same feelings of ownership and obligation towards his home, it came about that when it was attached by beast or man each male reacted identically and so produced unity of action in defense.

Doubtless more ages came and went before the idea of combination in defense produced its corollary, combination in attack. Again this unity of action was likelier due to change than reasoned thought. Stark famine drove men forth and chance presented them with some huge beast on which their separate hungers caused attack. Eventually, they may have been able to apply such combinations against beasts to dealings with their fellows, but it is more probable that some vagary of nature destroyed a cave and that its inmates driven by hunger to individual despair attacked a neighboring shelter all at once and so by lucky combination, won.

By some such steps was the value of combined action discovered until with the dawn of history we find; tribes, city states, and territorial principalities. Yet, nowhere are we told of the causes which had even then clearly set apart the leaders from the led.

Almost indubitably the cause was biological. In the family tribe the patriarch arrogated to himself, by virtue of his might and parenthood, the leadership. In the course of time, selective breeding, due to the appropriation by the chief of the most desirable females, produced some superior individuals among his progeny who, in their turn, gained eminence and repeated the process. In Egypt, and much later in Hawaii, such a system prevailed and produced a definite physical and conjoined mental superiority among the chiefly class.

Such superiority gave added opportunities for its own enhancement while at the same time reacting on the weaker masses to render them still less fit, by limiting their choice to imperfect mates and by depriving both them and their offspring of food either in quality or quantity comparable to that enjoyed by the chiefs.

In the combats undertaken by the tribes, individual vigor was the chief essential to success. This circumstance redounded to the advantage of the well nourished leaders who, by success at arms, not only improved their position, but also enhanced their reputation so that in both fact and theory they were respected as great. It is true that the opulence they acquired often sapped the virility which had procured it, but in the strata just below them were aspirants of equal breeding quick to take their place.

In these early combats the less well nourished of the commons died, while the survivors prospered through both plunder and chiefly favor, secured by association in success; hence finding war profitable they specialized in it and by its practice became the more adept.

Having now, as we believe, traced this evolution of the chief and his immediate followers, it seems pertinent to query what were the causes that still induced the weaker and less vigorous members of the population to fight.

As we have shown, the earlier fights arose from necessity and were persisted in through the pinch of hunger or the urge of lust; causes bearing with equal force upon every member of the community. Occasionally circumstances, or an able leader, caused success of a superior order so that general cupidity was added to the other causes of war. In all the ages during which such simple wars were going on, the chiefly class was developing and with it the habit of fighting under leaders; first spontaneously chosen, then selected for life, and finally born to the job.

Even with the early appearance of the hereditary chiefs, the optimism of youth (few savages grow old) and the unstable nature of society induced men to fight. They saw the returned soldiers richer and more important; while of those who failed at war they saw nothing and soon forgot their curtailed existence. Thus, optimism combined with habit to produce recruits while the carnal recompenses of victory and the cheapness of life produced hardihood in the actual fight.

Other causes evolved concurrently. Man and the lion possess in common the love for slaughtering the unresisting. It behooved men then either to be strong or else to seek the protection of those who were. For the privilege of existence, the unwarlike sold the birthright of independence. The more largely the weak lost equality, the more largely did its emoluments bulk in the perspective of their minds and the more strongly did the less base among them yearn for its possession. In strife these saw a door leading to the fulfillment of their desire.

Another cause for fighting found its source in the horrid monotony of pauperous existence. A certain amount of culture was, however, necessary before the mind could harbor this feeling for monotony, and could hardly have found place in those earlier times when each day's existence was but a dubious gamble with remorseless nature.

The situation at the dawn of history was then the result of the evolution of the hereditary chief supplied with soldiers whose incentives were, speaking generally, greed, the necessity for protection and habituated obedience; tinctured to a degree by ambition and spurred on to seek the romance of war by the prodding heel of impecunious monotony. While in the actual combat they were all held up by sex lusts, greed, and the mass instinct flowing from a multitude of communal emotions.

With the continued development of the intelligence, man evolved new names for old emotions while the brain children so conceived produced in their adolescence new conditions leading still more inevitably to war.

Since mating originated in violence, the females developed, or always possessed, a partiality for strength and ardor. This basic fact impinging on the mind through all its progress produced the notion that to be successful with the other sex man must possess might and violence. War was the natural place to demonstrate these traits. To disseminate the facts as proved, man had to appear well to his fellows so that they, in turn, would sing his praises to the shes. Thus was self respect, fear of fear, in a word courage (as a mental trait) evolved.

To fan the flame of this emotion came the bard, a dirty fellow probably, loath to fight, yet who found fat living by the simple feat of telling other men how great they were.

The courage above described is purely mental and hence requires conscious thought to make it operative. Its possession induces man to enter or seek danger, but it will not maintain him in its actual presence since there imminence of death stuns reason. Still, as the pride of valor is based on desire to appear well to others, the more conspicuous a man is the more is his pride sustained and buoyed up, the longer is he brave. This fact explains the invariably higher percent of casualties among leaders.

In addition to the above type of courage, there are two other sorts. First, the fortitude of experience as illustrated by old soldiers. This sort springs not from any particular virtue, but rather from the knowledge that battle is less dangerous than it seems. Second, the courage of the cornered rat, when either fear or rage obliterates intellect and the creature fights insensate in a blind effort to survive or slay.

The first of these two lesser classes is based on mental courage since clearly this was necessary, in the beginning, to get them to endure the scenes to which later habit made them casual. Since long association with fighters and military surroundings produces a sort of reflected image of this courage in the minds of men who have not, in fact, experienced the realities of war, and since in combat the nonchalance of veterans is imparted to recruits, the courage of experience is of vital military importance, and is the chief argument for professional armies.

The second sort of courage is purely individual and is the result of emotions not susceptible to prearranged stimulation, so is of no military value. It may be suggested that fear of punishment will arouse this animal courage. This is not so. To be influenced by the threat of punishment, man must retain his memory; this is inimical to the state of frenzy we have tried to describe.

Success, in any pursuit, but whets the appetite to greater desire, so the leaders who succeeded, and survived, yearned ever more strongly for war while with their increasing capacity their ambitions grew apace; instead of caverns, provinces became their goal. In like manner, but in lessening degree, the same feelings permeated the masses of their followers.

Before proceeding with the development of our subject, it is well to pause and here recapitulate the fundamental emotions which up to this point seem to have been proven the impulses which cause men to suffer wounds and death.

These are: hunger; sex and its simpler derivatives; unity of action, due to unity of impulses; biologically produced leaders; greed; the need for protection; ambition; romance; monotony and habit.

Not wholly unworthy, perhaps, of the ends they produce, yet, in their nakedness, devoid of those subtler emotions and roseate lights with which legend, art, and history have invested them.

We have adverted to the genius Bard of which Homer was among the earliest and most notorious example. But, it was to the institution of chivalry that the Bard, now a minstrel, owed his greatest eminence. Indeed, whether the minstrel was the child of chivalry and the grandchild of Christianity, or whether Christianity, through the bard, made chivalry, is a moot question. But the devices they two evolved, the minstrel and chivalry, to deck with wreaths the bloody hand, and raise the eminence the dripping lance, lived long after they themselves were spent. Truly, the knightly belt and golden spurs with the aura attaching to them have never been surpassed as a means of raising man above himself to deeds of selfless heroism. But we must not forget that broidered belt and gilded iron were but as worthless dross save when the eyes of lovely demoiselles flashed on those tinseled gauds, the glory of their age-old blandishments. And medals of today derive their potency from just that source—sex.

In addition to the effulgence which chivalry imparted to war, necessity and logic added to it utilitarian, but non-fundamental, restrictions and doctrines to which usage has assigned the force of law.

As illustrative of this last statement, attention is called to the fact that now care and respect for the wounded seems a noble and righteous act, whereas it is but the result of expediency and self interest. All might some day be wounded so the slaughter of the injured by the whole could well result in a similar fate to the slaughterers, when maimed. Help for the helpless springs from love of ourselves, not of our foes.

Fear of retaliation has in the same way modified the usage accorded enemy non-combatants. Also, experience showed that indiscriminate looting hindered more than it helped military operations and, in the case of a retreat over the ruined area, might well prove fatal to the over-thorough devastaters.

Long adherence to such conventions has now formed a mass conscience so that violations of such self-imposed rules would react profoundly against the violators. Witness the Lusitania.

Having noted the fundamental causes inciting the individual, and later the mass, to combat and having also examined to a degree the growth of sundry artificial incentives and restrictions incident to the waging of wars, we shall now investigate the agencies evolved to retain and stress the primitive emotions of individuals under circumstances which caused their spontaneous manifestation to become less and less a natural process.

As the everlasting strife went on, it became evident that certain individuals were more apt to it than were others. Also, as civilization improved, the power of resistance grew with it. More time was required to overrun a province or capture a town. The lengthening of operations led to a cooling of ardor; men were abundant who could take a day off to storm a cave, but were less numerous who could take a year off to capture a city.

Further, as the technique of killing improved, the fact developed that numerous unskilled amateurs were relatively less efficient and more costly than were fewer skilled professionals. As a result of this knowledge, leaders have at different epochs reduced their demands for quantity and satisfied themselves with quality.

When this project was first essayed it became patent that the naturally adept at war were insufficient in numbers to wholly replace unskilled hordes so the natural fighters were augmented by a certain number of the less efficient. In utilizing this system the further fact developed that something was lacking, particularly with the inapt portion of the force, this lack manifesting itself in diminished enthusiasm in the mass, arising from the nonexistence of that community of the sundry individual emotions formerly actuating its members.

The man who fights for a living must, unless he is a very rare person, live in order to profit by his fighting.

The fact that, from the beginning, valor has been the chief theme of song and story proves beyond question that at no time has its possession been a common attribute. This circumstance became strongly impressed on the leaders with the inception of semi-permanent forces and led to a quest for means to produce artificial traits whose result would simulate valor or replace the lack of the one-time individual emotions of lust and cupidity.

Eventually, this search led to the empirical discovery of habit. Means were adopted by which the incessant repetition, in peace training, of specific warlike acts produced so strong a habit, so stimulated, that is the automatic reflexes, that in combat the acts learned were performed subconsciously. But, like all hypnotic functions, the balance of this automatic state proved so finely adjusted that under sufficient stress, the hypnosis of habit (discipline) crumbled and men suddenly realizing their peril fled in as violent a manner as they had previously fought.

To counteract this tendency other means were evolved among which the most usually employed were: the strengthening of habit by still more rigorous rehearsals, the increased use of long service soldiers possessing the hardihood of experience, the infliction of savage punishments, the inflaming of men's minds with race and religious antipathies, the utilization of first local, then unit, and finally national patriotism.

Further, full advantage was at all times taken from a modified use of the allurements inherent in lust, cupidity, fame, and finally by the example of the few natural leaders and fighters whose acts, and the rewards conferred for them, aroused emulation.

In consideration of the foregoing, it appears that the habit forming repetitions called drills, and its various adjuncts called discipline and morale, are in effect but an attempt to produce a fictitious courage.

In order to give emphasis to our subsequent remarks, we repeat that the whole function of drill was to so fully impregnate men with the forms of combat, as practiced in training, that in battle they would still function. That the whole purpose of discipline and morale was to first induce men to go to war and, more important still, to so bolster up the habits formed by drill that in combat they would the longer endure.

Where the method was the phalanx, the phalanx was practiced with its ordered files and cadenced step. When the bow was in vogue, its use was rehearsed on foot or horseback according to the race and period.

Bowmen did not practice endlessly the evolutions of the hoplite nor mounted knights those of the legionary.

However, with the hunger for precedent awakened by the renaissance and stimulated by the printing press, a subtle change occurred. When Gustavus, Maurice, and Conde began to utilize small arms fire in ever increasing degree, the short range and low rate of their weapons made it vital that rigid linear formations be maintained so that the pikes and shot could mutually support each other and no opening be left wherein the hostile cavalry, hovering near, could insert itself.

With Frederick and his perfection of muzzle-loading fire the same causes, though in modified degree, still prevailed; rigidity was still essential. The successes he achieved and the notoriety attendant on them attracted many copyists so that though, in each succeeding war the necessities of the case grew ever less, his mechanism still prevailed until that luckless day when some pedant gave his system of close order seeming immortality by coining the phrase, Disciplinary Drills. Not only was this expression of itself alluring, but the inherent laziness of man seized upon it as a panacea for thought. The parrot like mastering of words of command, and tricks of execution, took little effort and no imagination. Phonographic drills boring to all concerned were executed in the soothing belief that by such acts the full duty of a soldier was accomplished.

So, today, with this procedure sanctified by years of usage, we find in our ceremonial formations and close order exercises simple copies of the battle formations and evolutions of the great Prussian, but without their raison d'etre [*sic*], for while with us they are utterly foreign to the battle-field, and may well become equally unadaptable to the march, in 1760 they were the key to victory.

We can but thank an ever just God that the person who invented Disciplinary Drills failed through ignorance or inadvertence to recognize the soul stirring efficacy, from the same point of view, of daily practice in the formation of the Roman tortuga. Had he been more erudite, oblong shields would still be an ordnance issue.

It seems pertinent to relate here, as illustrative of the utter folly with which man pursues means to the disregard of ends, that shortly after the death of Frederick a controversy arose among his officers as to the relative military value of a cadence of a hundred and twenty-two steps to the minute. In attempting to find a solution to this momentous question, many books were written and several duels occurred resulting in the death of three of the contraversionalists.

To return to our subject, it is maintained that these archaic drills in which we squander our time are worse than useless; they are actively harmful.

Battle is an orgy of disorder. No level lawns or marker flags exist to aid us while we strut ourselves in vain display, but rather, groups of weary wandering men seek gropingly for means to kill their foe. The sudden change from accustomed order to utter disorder—to chaos, but emphasizes the folly of schooling to precision and obedience where only fierceness and habituated disorder are useful.

We admit that in extended order we have Drills For Fighting but it is our experience, gained over a period of some twenty years, that the average officer who, as the word indicates, is the most numerous, will, due to his mediocre nature, and consequent lack of imagination and energy, spend at least four-fifths of his time in teaching formations and movements which have far less battle value than leap-frog. Nor is he wholly to blame, nor is the excellent officer immune from censure. Due to tradition, the measure most frequently applied in the determining of comparative excellence is the fictitious yardstick of precision drill. The good ones therefore strive to excel regardless of the futility of the means.

It is true that the theory of the necessity of pomp in war is ingrained in our nature, that we crave display and ceremony, in witness thereof note our countless uniformed marching clubs and societies. It would seem, however, that gala attire mingled in mass formation could satisfy this craving; even the orderly minds of ancient sculptors fail to present a Roman triumph with dressed ranks and ordered spears. Our present system disregards human nature in the infinite pains it takes to mingle

Prussian order with Quaker habiliments. Few of the brazen heroes adorning our village monuments have their clothing buttoned or their guns at a right shoulder. Disregarding this fact, we feed the craving for the gaudy and bizarre with doses of somber regularity.

We have already mentioned the fact that in addition to the use of endless repetition in Drills For Fighting, other agencies are necessary to keep man to his gruesome task in the terrible presence of death, and prevent him, particularly in his earlier experiences, from yielding to panic, which always hovers menacingly about even the best troops.

In considering these other means, we come upon certain disheartening tendencies resulting from increased culture.

For example, a survey of military punishments shows an ever diminishing severity caused chiefly by enhanced regard for the individual as such without consideration for the well being of his comrades. While as Christians, we should take comfort from this growth of leniency as an index of morality, and as citizens perhaps glory in it as an evidence of heightened respect for public opinion; as soldiers we must nonetheless admit that its existence makes the winning of battles ever more difficult.

In the World War, we had recourse to the stimulating force contained in hatred. Now while hatred has frequently proven very efficacious when founded on fact, the propaganda sort we attempted failed. We were poor vicarious haters, and had to rely rather on the mental tonic of a sense of duty. This stimulant becomes less and less potent as the enemy is approached because, due to its mental origin, it ceases to function in exact proportion to the shunting out of thought by the increasing imminence of death.

Better results would have been attained if the same public opinion which discouraged physical punishments had been enlisted to insure mental torture. In other words, had pride, a secondary sex emotion, been utilized. Unfortunately, undue consideration for individuals, fear of estranging potential voters, and a silly censorship prevented the people at home from ever knowing whom to honor and whom to blame. Few units would have failed to reach their objectives had their members been sure that the next morning the girls at home would have known. Unquestionably, individual injustice would have been done; such is the nature of war; but the doctrine of the Greatest Good would have justified such a course.

It is interesting to recall that during the Russo-Japanese War, a Japanese regiment which failed at 303 Meter Hill, if our memory serves us, was degraded, formed into a labor unit, and the facts promptly published.

Another impediment to leadership, and hence to success, is inherent in the obliteration of class enunciated in our constitution. However desirable this free and equal idea may be on political and social grounds, it is fraught with serious consequences insofar as the military is concerned.

Members of the gentle or lordly class to whom in peace respect is accorded by virtue of their birth, develop, by induction, a feeling of obligation to be worthy of this respect. Practically their only opportunity to demonstrate this worthiness comes to them in battle or in other times by grave emergency. Witness, for example, the almost universal heroism of the otherwise decadent French noblesse during the terror.

Conversely, the commoner continues to accord to the gentleman, when an officer, the same respect he previously accorded him as a civilian, and this, in its turn, tends to nourish and develop the feeling of leadership still further.

The following two incidents rather aptly illustrate the notion of hereditary superiority and consideration resulting from conditions such as above described.

On reaching London in the spring of 1917, we happened to converse with a withered little man who shared with us a seat on the Underground. In the course of the conversation he remarked that he, a clerk, had lost his two sons in the war, but immediately he added, "But, sir, that ain't nothin' as compared to our gentry. They were wonderful—they are all dead."

About a year later in France there was a long railway queue at a ticket window. Just ahead were two British soldiers. When we had been in line for some time a British subaltern came up and stepped into line ahead of the soldiers. Whereupon one of them said, "'E don't know no better, 'e ain't like our gentlemen hofficers wats dead."

With us and our mono-class system, the officer, particularly the new officer, has no inherent sense of superiority to sustain him and he is therefore either diffident, fawning, or else bolsters up his inferiority complex with undue harshness during training, while in battle he is more prone to forget his obligations and cease to lead.

On the other hand, the soldier seeing, as he often does, one of his sometime cronies made an officer has for him, initially, no feeling of respect, which fact in its turn reacts on the officer to lessen still further his self confidence.

Mobilization plans which contemplate officers and men for reserve units coming from the same localities are defective. The ex–ribbon clerk lieutenant from Mudville, may, for a time, be accepted at his shoulder strap valuation by the private from Swamp Hollow, but never by the men from home.

Of course, proven valor and ability evoke willing emulation and respect, but the proving takes time and opportunity.

The utility of patriotism, be it local, unit, or national, is admitted, but in our opinion not sufficiently explicated. Already, under mental punishment, we have referred to the powerful influence latent in local public opinion and evokable by a pitiless publicity as to successes and failures. Unit loyalty is most applicable to historic regiments and hence has but a limited application at the beginning of a war where we are dealing with the formation of amateur armies. It is clearly the reason behind the creation of Corps d'Elite. However, it quickly asserts itself; for among beardless veterans, life is vivid and very brief. In our Civil War such a feeling was clearly felt in and for such units as the Stone Wall brigade and the Bucktails, not to mention many others.

National patriotism seems at the present time a waning influence, barely discernible, as a sort of mirage of hot air against the pale pink sunset of masculine virility. This subsidence is partly traceable to socialistic propaganda but chiefly takes its source from the fact that in huge nations local interests supervene to eliminate, either wholly or in part, a just appreciation of national matters. In our own case, this condition is further aggravated by lack of race homogeneity.

Having sketched the sources from which the fighting spirit takes its origin and having further depicted certain of the present day drawbacks to the full utilization of these primary springs, we shall, in closing, attempt to advance some remedies purposed to correct the defects in our system in preparation for that resumption of war which the inevitable cycle of history unmistakably proclaims.

Insofar as the technique of Drills For Fighting is concerned, its nature is too complex for discussion within the space of this article. We, therefore, dismiss it for the present with a simple assertion as to its vital nature. In a subsequent article, we purpose to extend and examine the subject in greater detail.

In considering the moral stimulants with which we propose to augment our technique, the first in order is punishment.

Due to maudlin sentimentality it is not possible to cause the mass of a nation to view military punishments from their proper angle, namely as administrative rather than judicial acts, whose purpose in wartime is not to wreak vengeance on the guilty, but, rather, to restrain the innocent.

For example, the idea behind the death sentence for such acts as desertion, sleeping on post, skulking, etc., is not inherent in the offense itself. Desertion has, perhaps, no extenuation but the other offenses usually have.

The poor tired boy who sleeps on his post is more to be pitied than blamed insofar as his individual case is concerned. But the act, harmless in itself, may have exposed scores or hundreds of his equally deserving comrades to capture, wounds, and death. It is for their sakes, not his fault, that the final penalty should be exacted of him.

A man may, under the influence of fatigue, so forget his obligations as to chance a term of imprisonment against a moment's oblivion, but he will think several times before he makes the same gamble with the certainty of death; his own death, not that of his comrades.

So with the skulker; the act in itself may be the natural outgrowth of lifelong teachings in safety first, may arise from the instinct of self preservation, may be the result of nervous collapse caused by fatigue, or may be sheer lack of guts. None of these reasons is in itself very abnormal. Nor, when viewed from the then frame of mind of the skulker, is the resulting act very heinous. Only when we consider the act apart from its results does its enormity become apparent. In the first place, the skulker jeopardizes and may negate the efforts and sacrifices of many gallant comrades. In the second place, he sets an example, which spreading with the rapidity of a prairie fire, sweeps others of his kidney to acts of similar baseness. Worst of all, he cuts the very root of military virtue, which is based on mutual confidence.

The execution of the skulker is necessary, not for his sin, but for his betrayal of his comrades. Judas is execrated for the betrayal of One, should he who betrays hundreds escape?

The man who shirks does so from fear of wounds or death, seldom in actuality, measurable by odds greater than one to five. If he can be assured that the next day his odds will have changed to one hundred to nothing, with the chance of wounds eliminated, he will be more chary in his shirking.

A long war, particularly if it is initially unsuccessful, may in the end convince people of the expediency of these views. In the beginning, much inertia must be overcome. The press and public men could aid in forming opinion to support the military—that they will so aid is more than doubtful. So, we are faced with the fact that, in the beginning, our men will skulk and sleep in the usual proportions until the habit of battle and the stern measures adopted by their afflicted comrades enforces some check to their predilections.

Of course a check always exists in the application of those preventive measures euphemistically called Battle Discipline which some few officers and noncoms have the courage to use.

The only immediate and practicable remedy lies in so modified a censorship that a Pitiless Publicity shall expose with equal promptness the doings of the hero and the knave to that most merciless of tribunals—home town gossip. The effects of this device would be most far reaching since the ease and promptness of communication have made available to the soldier the whip lash of home opinion in a degree never before approximated.

Men do not fight for pay—they must have pride. Any system which deprives them of glory while rendering them immune to scorn is absurd.

To maintain the sequence of our previous remarks, we shall next investigate the stimulating power of hatred.

The word is over strong to express those race and cultural differences which the weakening tendencies of modern civilization still permit us to use. Yet, degenerate as they are, they are still worthy of consideration. The best means of emphasizing them is to use racial differences to stimulate the superiority complex. For example, there is no physiological reason why a diet of frog legs should be less manly than one of cows' ribs, yet the constant allusion to their enemies of a hundred years ago as frog eaters undoubtedly inspired the British of that time with a contempt for them. Similarly, a difference of opinion as to sartorial trimming has been used by both sides to bolster up self esteem to the prejudice of their adversaries.

Since the insertion here of similar pertinent remarks as to the gastronomic or other peculiarities of our potential opponents might well unbalance the peace of the world, we shall refrain; but a little forethought could, we feel sure, provide us with many slurring epithets.

In addition to these puerile, but useful, means of developing self esteem, there is the universal device latent in atrocities. Good atrocities are easy to invent and difficult to refute, since in all argument the assertive has it over the explanative. In picking such heinous acts it is best to choose those we attribute to the enemy from among the ones of which we ourselves are the most apt to be guilty, since, if we are caught, we can then explain our acts to be retaliations. On the other hand, since we have never been blatantly guilty of acts against religion, unless it differed from our own, or against women or children, much capital can be made by instantly accusing the enemy of being ungodly and ungallant.

Finally, in the event of wars waged in our own territory, we have the hatred coeval with the race which now, as in the beginning, has ever existed against the attacker of the home.

Under punishments, we have already mentioned the utility of local pride or patriotism, but its potentialities can be still further exploited.

Nations which have created regimental depots have done so with this object in view.

To obtain the best results, recruitment should be done from the area contiguous to the depot and, at it, all recruits should receive their initial training from detachments of the unit they are to join.

At the same time all convalescent wounded should be passed through the depot on their way back to the front. By these means, coupled with great candor and promptitude in the reporting of the current successes and failures of the regiment, civilian interest will be aroused. Due to the presence of the returning veterans, the history, exploits, and traditions of the unit will be earlier imparted to the new soldiers. Further, since veterans are never averse to enhancing their own reputation by stories, true or otherwise, of their recent exploits and, while warmed by the ardor of their recitals, they stress only the glory and excitement of combat, these stories give to the recruit pictures of war which, while illusory, are permanent. So stimulated, the young soldier determines to emulate or surpass his predecessors and having frequently so announced is later deterred from back sliding by the fear of local disgrace with its attendant loss of standing among the fair sex. He thus cultivates in himself prospective hardihood.

Another trait which is played up by the Unit Depot system is the latent desire in all men for posthumous celebrity. For, birth control to the contrary notwithstanding, all men possess that feeling so well illustrated by the story of the soldier who on being called before his captain for fighting a comrade excused himself by saying, "Well sir, it was this way. I was cleaning the latrine and this guy comes by and says to me, 'What are you going to tell your children when they ask you what you did in the great war?' So I hit him."

With the depot system it will be easier for the children to know.

Unit pride is but an extension of the local spirit just described and is subject to the same influences. In addition, it is fundamental to its existence that absolute permanence from both officers and men must be maintained. No claims of economy or expediency can ever be justified where they involve incessant shifting of men from unit to unit until they have no more lares and penates than a traveling salesman.

It is noteworthy that in our present general mobilization plans, a replacement for a Texas division may well come from Boston.

With the Regular Army this system could be inaugurated in peacetime, for when the exigencies of the service demand that Captain John Doe be detailed in a staff corps or department he should carry with him his identity by the use of the title and insignia of Captain John Doe, Nth Infantry, Q.M.C., etc. The formation of such a system would require some bookkeeping but with the restrictions as to the number of battalions in a regiment removed, mobilization would always find more than enough places in the old regimental home for all its wandering children.

While the following remarks are probably as unattainable as Moore's famed republic, we cannot refrain from inserting them. We believe that much better results would be attained if the present pompous and empty Organized Reserve divisions, brigades, and regiments were scrapped and the units of that force were limited to battalions; while the battalions, themselves, should be made integral with existing regular regiments as 4th, 5th, etc., Battalions, Nth Infantry. As a prerequisite for such an arrangement the regular regiments should be localized for recruiting and the subjoined Organized Reserve battalions be given identical localities.

Two immediate advantages would accrue from such an arrangement: first, the elimination of Reserve Officers with rank out of all comparison with their experience and abilities; second, a greater stimulus to local civilian interest in both regular and reserve units.

Such a step clearly presupposes the adoption of the so-called British system. It should be recalled, however, that prior to 1898 it was in large measure our own. A further discussion of this subject from a tactical view point will appear in a subsequent paper on Drills and Fighting.

As has already been pointed out, our present bulk makes active National Patriotism largely impalpable. Stress of war, particularly if one of invasion, or one in which we were initially unsuccessful will, to a degree, crystallize this feeling, but in the meantime the situation would be improved if pacifist moves were treated with less tolerance. The prompt shooting of some scores of conscientious objectors would go far towards removing bellicose inhibitions.

Another move which might be easily instituted at the beginning of the next war would be to explain to editors, most of whom are patriotic, that the printing of sob stories amounts to a traitorous act. Men in combat are too weary or excited to entertain the thoughts and emotions attributed to soldiers by the sick brains of unwarlike writers while the pre-battle reading of such stuff by new soldiers simply subjects them to useless and brutal mental torture, as they anticipate feeling these fictitious emotions and, in prospect, suffer many pangs which their actual experience will subsequently prove nonexistent.

In our initial catalogue of inciting emotions we enumerated lust and greed. But today improved culture steps in again to deprive us of the major part of their influence. The sacking of places taken by assault was the chief means of pandering to these crude feelings. Since our civilization cannot now stomach such acts we are forced to abandon them. Certain conditions may, however, yet arise where circumstances of great hardship can be exploited to inspire our men with the hope that success will grant them full bellies and warm clothes. The battle of Gettysburg was precipitated by the rumor that the town contained a large supply of shoes.

The popular expression, "Soldier Boys," has more truth in it than its poetic originator probably guessed. The fighting ranks (or armies) are largely composed of boys, and the simplicity of the life led by men in campaign tends to retain and redevelop their boyish propensities. While the theory, that in passing from the germ plasm to the grave we relive in a brief span the whole gamut of our evolutionary existence, may not be wholly tenable, it is nonetheless certain that the boy is more nearly similar to uncivilized man than is the person of greater age.

For this reason soldiers respond very readily to the simple emotional stimulants which formerly actuated the race. Of these the most potent was the sex originated desire to appear well, to be a hell of a fellow. We see this desire for self expression and laudation evinced in the bizarre costumes, antics, and hair cuts of school and college boys. Putting men in uniform, usually ill-fitting, not only deprives the youth of his power of self expression, but, to a degree, hurts his pride by making him look like his associates.

In the gay uniform given by Napoleon to the Young Guard we find him realizing and profiting by this fact.

The fortune of many Semitic tradesmen was started in 1919 by the sale to returning Heroes of gaudy, but unauthorized and meaningless Campaign Ribbons, bought with no intent to deceive, but purely to please the ladies.

Every recurring election or club convention depicts the same blossoming out of the male population in badges and bloomers.

In the army, the ribbon has replaced the knightly spur and belt and at a greatly reduced cost.

Its possession gives differentiation, distinction, and fame. For the privilege of wearing a dime's worth of taffeta, a man will do deeds which all the treasure of the Incas were impotent to cause him to attempt.

In the World War we increased our decorations from one to three, but it was not enough. And, the parsimony and delay attendant on their distribution took from them much of the effect they were intended to produce. Fearful that one unworthy might be decorated, we examined, hesitated, and hectored our heroes; utterly forgetful of the fact that a coward dressed as a brave man will change from his cowardice and, in nine cases out of ten, will on the next occasion demonstrate the qualities fortuitously emblazoned on his chest.

We must have more decorations and we must give them with no niggard hand. The story of the young soldier who, on being asked by Napoleon what he desired in recompense for an heroic act said, "The Legion of Honor, Sire," and when the Emperor replied, "My boy, you are over young for such an honor," again answered, "Sire, in your service we do not grow old," is as true as it is tragic. Our men will not grow old. We must exploit their abilities and satisfy their longings to the uttermost during the brief span of their existence. Surely a machine gun nest for an inch of satin is a bargain not to be lightly passed up.

The frame of mind which places the invariably unsuccessful attempts to delude the enemy above the inspirational influence of a distinctive uniform is to us incomprehensible. If a chasseur's cap could make a Blue Devil out of a peasant, think what a pink feather could make of our men!

In addition to the present battle honors on the regimental colors, the colors of specially distinguished units should be exempted from saluting while passing in review. Nice distinctions might be made in this privilege, as for example, allowing some only to salute major generals, some more distinguished only lieutenant generals, some generals, and some, the highest, only the president.

Endless other expedients could be enumerated, all tending to produce distinctions based on military merit, to emphasize valor, to vividly proclaim the super-fighter.

We know that we will be told that Americans do not care for such things, that the supply problems will be made more difficult, that injustices will be done. To such remarks we reply that the Spirit of the Soldier Boy is ever the same, ever the simple, vain, ingenuous savage soul of youth; that supplies difficulties seldom trouble victors, and that war is only resorted to when justice has failed.

You can rely on punishment, habit, and the valor of experience with long service professional armies, such as we shall never have. With amateur armies you must lead by seduction. War may be hell, but for John Doughboy there is a heaven of suggestion in anticipating what Annie Rooney will say when she sees him in his pink feather and his new medal.

"In war," said the Emperor, "men are nothing, a man is everything." All history vindicates the remark. The subterfuges and stimulants we have mentioned arise from and owe their existence to the lack of men of the leader type and all of them are but auxiliary means to which the leader spirit gives life and utility. Mind you, we are not concerned here with the great military luminaries. So far as we can discern, this select group must be classed as biological incidents whose existence is due to the fortuitous blending of complementary blood lines at epochs where chance or destiny intervenes to give scope to their peculiar abilities. What we must acquire to lead our men are the lesser combat chiefs. In this country we start our search handicapped by the absence of potential leaders consequent upon our lack of civilian class distinction. This condition cannot be altered, but other means are at hand among which are the following.

By increasing greatly the number of graduates from the Military Academy and discharging the surplus after a year's service with troops, we will secure a certain number of men who, by education and training, are better fitted for company officers than would be the same individuals fresh from the counter or the farm and minus the ingrained traditions and attached prestige consequent to graduation from that great school.

Additional material is also being accumulated from graduates of the several R.O.T.C. Colleges. Such men have an initial prestige which will help bridge the gap until a reputation of demonstrated ability is secured in battle.

The limited number of noncommissioned officers with Reserve commissions are superior to the college men for while they lack education, they vastly surpass them in experience and assurance.

Reserve corps officers with World War experience are fast becoming useless because, in company grades they are too old, while for field grades, with very few exceptions, they are utterly useless. We base this assertion on the fact that the qualifications for successful leadership in business and war are similar. Due to high competition in business, few successful men in that walk of life can find time to devote to military study and none of them can, in the brief space of summer training, find time or opportunity to master minutiae of war, or attain the habit of command on which military leadership is founded.

Even if all the above sources of supply are developed to the full, the next war will almost surely find us facing a dearth of officers and we will be tempted to have recourse to Training Camps. Now, in spite of our vaunted democracy, there is nonetheless a certain tendency to class distinction based on educational qualifications. The dividing line seeming to sharply separate those who hold and those who do not possess a college diploma. In our experience, this distinction is illusory so far as command ability is concerned. Murat and Villa could never have entered a training camp.

The notion that military ability takes its source in high-powered thinking is very congenial to us, nicking as it does with the national mania for superficial education. The theory is further stimulated by historians, who, being students themselves, are inclined to depict their heroes as being highly endowed with the same traits which in themselves they so greatly reverence. The chance reference to 1870 as the Schoolmasters' War is often quoted by them with complete disregard of the fact that is initial successes were due to the ferocity of the unscholastic Steinmetz. It may well be that the greatest soldiers have possessed superior intellects, may have been thinkers; but this was not their dominant characteristic. With the possible exception of Moltke, all great generals with whom we are familiar owed their success to indomitable wills and tremendous energy in execution and they achieved their initial hold upon the hearts of their troops by acts of demonstrated valor. However, we digress; the great leaders are not our responsibility, but God's.

In the lower grades in a great war special education will be impossible and general education useless. We must commission only brave and energetic men. As the only means of carrying out this notion, it is submitted that there be no training camps, that all commissions after the first battle go to soldiers of proven combat ability, and that all replacements be in the grade of private. In making this assertion it is evident that we are abandoning our thesis as to the advantages derivable from hereditary class distinctions. Such is exactly the case. It is futile to consider conditions which for us cannot exist. What we must do is to go back a thousand years or so and reconstitute in our armies the aristocracy of valor in which all aristocracy originated.

Men commissioned and promoted in accordance with this plan will have, as a start, the aura of proven courage; they will have further the prestige of battle experience. Promotion so gained will arouse in them pride and a sense of obligation to be worthy of the honors conferred. Finally, we shall very rapidly develop a hierarchy of courage and infinite solidarity since the junior will owe to himself his ability and to his superior its recognition; for captured objectives, not mildewed diplomas, will mark the road to preferment. If confirmation for these remarks is necessary, it exists in two statement made by the master of war, Napoleon. "Better," he said, "an army of stags lead by a lion than army of lions led by a stag." And again, "Every French soldier carries in his knapsack the baton of a Marshall of France."

Having completed our investigation, enumerated our difficulties, and advanced ideas for their correction, we close by summarizing the result of our efforts thus.

Nothing new was discovered since the soul of man is changeless.

Our difficulties differ in manifestation but not in nature from those Alexander experienced and Caesar knew.

Our success or failure in the next war will depend on our ability to face the naked facts as they exist, and to utilize our means not as we would, but as we may.

THREE

Technology and War

Throughout military history, there have been seemingly endless moves and countermoves in the fields of weapons technology and tactical employment—and in the interplay between the two. In this highly competitive strategic environment, there is no shortage of creativity and guile in devising new technical or tactical means of achieving battlefield victory. In recent decades, the world has witnessed a rapid advance in weapons technologies such as precision-guided munitions, low-observable technologies, unmanned aerial vehicles, and computer networks as well as a return to remarkably old-fashioned forms of tribal, political, and religious violence often fought with an ad hoc mix of improvised and archaic weapons. Indeed, the adoption of low-intensity and hybrid forms of warfare by many of America's adversaries can be seen as a tacit admission of the futility of fighting the United States on its own terms and a clever means for seeking victory.

In the November 1930 *Cavalry Journal*, Patton published "The Effect of Weapons on War," which expressed his thoughts on the subject. Warfare, the article declares, has been in a perpetual state of flux between the development of new weapons technology and tactics and responses to those innovations. The interplay between technology and tactics is an enduring and evolving problem that will never be fully "solved" but will continually manifest itself in new iterations of an ancient pattern.

The interaction between technical innovation and tactical response fascinated Patton. He believed it was necessary to study historical patterns in order to anticipate the future direction of war and remain one step ahead of potential rivals. Patton read a massive number of military histories and compiled detailed notes about the technical and tactical keys to victory in each era he studied. To make his academic studies more tangible, he amassed a large collection of arms and armor and was a true expert on antique swords and firearms and their employment.[1]

In addition, Patton was very hands-on in his understanding of technical problems of the day. From an early age, motorized vehicles fascinated him, and he was a proficient mechanic and tinkerer. When Patton commanded tank units during World War I and the interwar period, he set about understanding the mechanical workings of his machines with an uncommon vigor, quickly mastering the technical complexities of these new weapons. When creating tactics for the embryonic American tank force, Patton was able to apply his unique combination of historical knowledge and technical skill to create doctrinal solutions to a wide range of previously unsolved problems, such as how to perform refueling and vehicle recovery under fire.[2]

For Patton, strategic uncertainly was omnipresent in history, but the effects of these disruptions could be mitigated by careful study and an appreciation for the value of the individual. If properly led, good troops had the ability to solve problems and achieve victory. Patton understood that in the end, the drive and initiative of troops, not new weapons systems, made the biggest difference on the battlefield. He believed that neither technology nor innovation alone was a substitute for human imitative and courage, and he incorporated this belief into his dynamic leadership style.

This article provides an insightful view into Patton's deep knowledge of historical and technical issues as well as his philosophy of leadership and war. In it, he claims that the basic nature of war has remained unaltered, despite the newest methods of warfare. For Patton, war is about killing people and breaking things. Wars are won by people, not machines. Although the technical issues of the present have changed considerably since he wrote this article, the need for leadership and sacrifice has not. Despite the technological advances of recent decades, this article is a sobering and enduring reminder that war is an intensely personal and human affair.

THE EFFECT OF WEAPONS ON WAR

Cavalry Journal, November 1930

When Samson took the fresh jawbone of an ass and slew a thousand men therewith, he probably started such a vogue for the weapon, particularly among the Philistines, that for years no prudent donkey dared to bray. Yet, despite its initial popularity it was discarded and now appears only as a barrage instrument in acrimonious debate.

Turning from sacred to profane history, we find it replete with similar instances of military instruments, each in its day heralded as the "dernier cri," the key to victory. Yet, each in its turn retiring to its proper place of useful, though not spectacular, importance.

Of yore, the chariot, the elephant, armor of various sorts, Greek fire, the longbow, and gunpowder, to mention only a few, were each acclaimed. Within our memory the dynamite gun and the submarine were similarly lauded. Today, the tank, gas, and the airplane are aspirants for a place on the list.

In investigating the question, let us begin by picturing, if we may, the cataclysmic effect produced on primordial society by the first savage who chanced to use a splintered rib as a means of giving point to his demands for a larger share of meat and women. How they gibbered around the half gnawed bison as with signs and gutturals they described the fight.

How their hairy bellies palpitated as into the twilight of their minds the idea flickered that they, too, might be so struck. "Romance is dead," they growled. "The day of tooth and fingernail is done."

Eons perchance rolled by before some timorous soul, fleeing in vain the questing menace of a prodding point, seized, in his agony of terror, a jagged stone and, squealing as he hurled it, saw the pikeman fall. Trembling, he knew that artillery was born. Continuing, it is easy to imagine the appearance of a wattled shield to fend off the stone and after the inevitable lag phase, ages long when men thought dimly, such shields, in turn, made useless by the sling and throwing stick. Another lag and then the bullhide shield restored the balance and robbed the sling and javelin of their lead. Consider how the scythe chariots were rendered innocuous by the simple means of opening the ranks to let them rattle through. Later, at Zama, similar tactics permitted Scipio to render futile the tankish charge of Hannibal's elephants; no longer a novelty and so dreaded as when Phyrrus used them. Again, consider how, off Sicily the Roman Ravens (boarding bridges) confounded and destroyed the far superior Carthaginian fleet; not by their inherent value, but by their devastating effect of their novelty. They, too, quickly passed.

The long struggle between armor and weapons abounds in like examples of alternating successes. When Cortez defeated an army by a charge of fourteen horses, it was not the valor of his "caballeros," but the fear induced by the novelty of their mounts, which routed the Indians. In this case, however, the results attained are not traceable wholly to surprise. The rush of horsemen, and similarly of tanks, reawakens a submerged race memory of ancient flights before the devastating rush of long extinct carnivora. We might continue almost without limit eliciting further examples, but repetition is wearisome and enough has been said to justify us in formulating an axiom. It is; the initial appearance of each new weapon or military device has ever marked the zenith of its tactical effect, though usually the nadir of its technical efficiency.

Surprise is the most ancient and most potent of military methods. Novelty is a form of surprise, and it is surprise (the fear of the unknown), not power, which appalls us.

The wrestling adage that there is a block for every hold applies equally to war. Each new device is invariably followed by its self induced counter. The utilization of these new methods and their counters, these holds and blocks, is highly useful in that they add to our combat repertoire. But their employment is fraught with danger, if, beguiled by their transitory preeminence, we place our reliance wholly upon them.

It is only in the writings of the romantic novelists that we find the hero successful through the knowledge of some secret lunge. In the duel or in the fencing room, success goes to the man of many good attacks and sound parries; to the man who uses all of the means at hand for the accomplishment of the end sought, victory.

Here it is well to pause a moment to examine certain characteristics which have definitely marked the march of military evolution. From the very beginning, our gifted species has expended vast amounts of time and ingenuity in a strenuous, though futile, effort to devise safe methods of war; means of killing without being killed. Ardant du Picq sums it very aptly when he says, "Man engages in battle for the purpose of gaining victory, not for the purpose of fighting."

Defensive devices are an outgrowth of the same desire; the stone and the shield, the lance and armor, gas and the mask. Obviously the emotion back of these manifestations is love of life; an emotion which from age to age has grown stronger as the chances for its enjoyment have increased.

The hero is of truth a rarity. The most striking proof of this is found in the fact that throughout myth, legend, song, and story he has invariably shared with that other rarity, beauty, the place preeminent. Much heroism exists, but few heroes. It is rather disheartening to observe that man in his efforts to reduce danger has enhanced the requisites for courage necessary to withstand it. The sweat, noise, excitement, and bodily contact of the close encounter act as a sedative on the brain, the seat of fear. After the rush has started it takes less hardihood to charge than to sit stolidly in a ditch awaiting dissolution via the impersonal belch of a dropping shell.

In attempting to assign just valuations to the latest lethal devices, we shall not go far wrong if we keep in mind the lessons of history. In the first place, living in a mechanical age, we are prone to exaggerate the value of machines. Again, lay opinion is chiefly formed by the press, where novelty is always "front page stuff." Erroneous habits of thought also play a part. During the World War, correspondents were not allowed at the extreme front where the actual bludgeoning of war took place. Necessity imposed on them the task of making copy of what they saw; guns and machines, mostly; hence it happened that they put undue emphasis on these elements and so formed in the minds of their readers a habit of reverence for machines.

The romantic literature of the war, now as always, centers on the exploits of heroes. Unthinking people imagine that in the future all machines will be operated by these rare individuals and that the phenomenal results attained by the few will be duplicated by the many. In sport we have Sande, Tilden, and Jones, whose exceptional capabilities we admit and admire. Yet, in war we fondly imagine whole armies of Sergeant Yorks and Guynemers. Popular antipathy to unhappy endings induces writers to have their heroes "live happily ever after," whereas, in fact, only too many citations for valor end, "For this act he was awarded a Medal of Honor, posthumous."

The use of gas as a weapon is abhorred by most civilized nations. Those who in future first resort to it may well find themselves condemned by public opinion. In short, it is against the rules. But, will such rules, such scraps of paper, deter belligerents? We fear not. When two highly paid athletes contend for honors in the squared circle they too are bound by rules; so much so in fact that of late rules have proven more potent than blows. War is not a contest with gloves. It is resorted to only when laws (which are rules) have failed. If some adversary gasses us, we can under the rules, gas him. Hence, it is not brutal, but merely intelligent, to investigate the probable future military effects of gas.

What are we to expect? Casualties, certainly; destruction, no. Gas is no more devastating to the prepared soldier than were stones to the shield guarded barbarian. It is a powerful and effective weapon, but the day of its omnipotence and the day of its birth were one. The gruesome pictures of whole populations writhing in their last agonies amidst the fumes of an all destroying vapor, are "bunk."

Setting aside the chemical difficulties and mechanical complications inherent to such an act, we have a much stronger and simpler reason for this conclusion. For centuries all wounded and such unwounded prisoners as were valueless as slaves had their throats cut. No one was shocked; it was the custom. Finally, it occurred to some altruistic and thoughtful soldier that while the practice was excellent so long as he was the victor, it had its drawbacks in the not unlikely event of his being the vanquished. The notion of humane treatment for the foe was born. Years of use sanctified the idea; it became the custom. Yet, the horrid thought pops up that help for the helpless sprang from love of ourselves, not of others; from fear of retaliation. The same situation effects the noisome idea of gassing noncombatants. It is contrary to our developed sensibilities, it will produce retaliations; it is not a safe method of war.

Shortly after the Spanish War Colonel T. R. Roosevelt wrote a book called "The War in Cuba." Mr. Dooley, in discoursing on it said, "I have but one suggestion to offer the Colonel. He should have called his book, 'Alone in Cuba.'"

The same remark might justly be applied to those who now proclaim that the airplane should be the sole means of waging future wars. They think that they will be alone in the air. So far as a major contest is concerned, this notion is absurd. The enemy will be there, too, and it will be a case of dog eat dog. When planes attacked us in France, we hid and prayed; now we shoot back and with an ever increasing effect. There is an old saying in the army that no pursuit is so hot as that of an unresisting foe. When the foe fights back, ardor slackens. Have you ever noticed the fervent manner in which a terrier chases a cat until the cat turns? Then how often he remembers that he has an immediate engagement elsewhere.

Air attacks will be numerous and bloody; such is the nature of combat. They will be no more conclusive than are the independent attacks of any of the other arms. As for bombing raids against cities, London still stands, and the inevitability of reprisals will tend to reduce still more this messy business. The airplane is here to stay. It is a great arm, but it has no more replaced all others than did gunpowder.

That fecund mother, Necessity, who at Troy produced the wooden horse, begot of the machine gun that horse's modern prototype, the tank; an identical twin to all of her preceding military offspring; the counter to the latest form of defense.

At first the tank, despite innumerable ills of childhood, enhanced in this case by premature birth, was a success. It was a surprise. As it waxed stronger it still prevailed, to a degree, due to its inherent worth. It has been likened to an armored knight. The first emblem of our tank corps was such a warrior. The similarity is too apt. So long as the knight combined movement with invulnerability he prospered. When he sacrificed mobility for protection, he passed on.

In the World War, infantry with their machine guns were impotent against tanks. Only direct hits by artillery, bad going, and above all, engine trouble, stopped tanks. Now every arm has its quota of antitank weapons which are quite effective. The terror of surprise is gone. In a major war, tanks will fight tanks. A land Trafalgar will be brief, bloody, and pyrrhic in its results.

By land and sea it is the same old story of guns and armor. We shall always have battleships, and we shall always have tanks and land destroyers, too, in the form of armored cars. Also we shall have losses. Utopia is not yet. The tank is vastly potent and rigorously limited; it is not and never has been a life insurance policy for tank gunners and drivers. It has no more the power to replace the other arms than had the long bow.

General Forrest said, "War means fighting and fighting means killing." When that grim time comes again, remember that all arms are potent, none is paramount.

We are always well aware that our efforts to prove the fallibility of weapons as a key to victory are wasted on students of history. Unfortunately, the lure of the bizarre tends to make mankind as a whole disregard its teachings. Nor is this a phenomenon confined only to things military. When sages point to the sublime inevitability of the cycles of history in morals, politics, dress, and so on, they are told, "True for you, but things have changed. We have the radio now and women vote." Similarly in matters military when we point to the endless cycle of holds and blocks we are told, "That was all true in the days of Napoleon, but now we have gas, tanks, airplanes, or what you will."

So far as we know, few, if any, victories are traceable to weapons.

Caesar destroyed the poorly armed Gauls and he did the same to the armed Legions of Pompeii.

In 1866, Prussia defeated the less well armed Austrians; in 1870, she destroyed the better armed French.

Advertisements to the contrary notwithstanding, Big Business does not owe its bigness to a filing system (a business weapon).

Already in this article we have made use of part of Napoleon's magnificent definition of genius. Here it is in full. He says, "Genius is the ability to utilize all the means at hand for the accomplishment of the end sought."

The thought applies equally to weapons. We must use them all. To us it seems that those persons who would scrap the old and rely only on the new are on a mental parity with the poor man who pawns his shirt and trousers to buy an overcoat, only to find that it is burdensome in summer and not wholly satisfying even in January. Wars are fought with men, not weapons. It is the spirit of the men who fight, and the spirit of the men who lead, which gains the victory. In biblical times this spirit was ascribed, probably rightly, to the Lord. It was the Spirit of the Lord, courage, which came mightily upon Samson at Lehi that gained the victory. It was not the jawbone of an ass.

FOUR

Patterns of Success

One of the most enduring questions in strategic studies concerns the role that individual leaders, particularly general officers, play in altering the course of war. Although it appears obvious that leadership matters in shaping outcomes, the exact impact of any leader in determining the outcome of a battle or a war is difficult to measure. This debate divides those who believe that leadership is less important than factors such as technology, mass, surprise, and so on from those who believe that dynamic leadership is the greatest determinant to success in war. Given the complexity of the question and the deeply entrenched beliefs of both sides, it is unlikely that either side in this debate will ever fully prevail. Despite the deadlock, this discussion is valuable because it forces students of conflict to ask key questions about the nature of conflict and make better arguments about the relative importance of different factors in producing victory across time.[1]

Patton was fascinated by these questions and studied them in great detail. In his January 1931 *Cavalry Journal* article, "Success in War," he returned to the subject of leadership as the key determinant of victory. As in the article in chapter 3, Patton skillfully combines his deep historical knowledge with his personal observations and refines his assertion that inspirational leadership is the single greatest determinant of success or failure.

Although factors such as numbers, intelligence, planning, technology, and initiative are necessary for victory, Patton believed that none of them are sufficient for victory without leadership. The recently defeated Imperial German Army was an illustrative case study. The Germans had lost World War I, Patton asserted, not because of a failure of preparation or tactical precision but because they lacked effective leadership to execute their plans. Despite the vaunted reputation of the German army for professionalism, its inability to achieve victory was a warning to Patton not to ignore the human dimension of the military arts.

Because of its critical importance, leadership must be thoroughly studied and carefully nurtured, Patton believed. Successful leadership was not a product of a fixed or easily replicable formula, he argued, but a force of nature in its own right that was omnipresent in warfare and the human experience in general. Patton thought that leadership had evaded systematic study because it was difficult to quantify or define. Although it is comparatively easy to observe, leadership is essentially nonlinear in nature and frustratingly difficult to describe or pass on to others. This problem is confounded by distortions in the historical record inherent in self-promotion, the difficulty of assigning causation, faulty memories, and hero worship. Yet Patton was insistent that leadership mattered. Throughout his career, he reiterated this point, observing that "*leadership* is the thing that wins battles. I have it, but I'll be *damned if I can define it.*"[2]

Echoing Prussian philosopher of war Carl von Clausewitz, Patton noted that war on paper was different from actual war and that a wide range of intervening variables such as "hunger, emotion, personality, fatigue, leadership, and many other imponderable yet vital factors" combine to shape war's outcome. Much like Clausewitz, Patton believed that military genius could help mitigate the effects of friction and chaos and could often compensate for deficiencies in other key areas, such as numbers and intelligence.[3]

Patton believed that he had an innate military genius and that true warriors were often born, not made. He cultivated his image as the perfect natural warrior and convinced many that he had a uniquely atavistic warrior soul.[4] Patton frequently referenced his ancestors, some of whom had fought bravely in the American Civil War, and claimed that he was the reincarnation of soldiers who had fought in countless battles across the centuries. In the opening stanza to his most famous poem, "Through a Glass Darkly," Patton references this belief:

> *Through the travail of the ages*
> *Midst the pomp and toil of war*
> *Have I fought and strove and perished*
> *Countless times upon this star.*[5]

Despite a confidence in his own unique military destiny, Patton also believed in the power of self-improvement, that soldiers could improve their leadership skills through study and practice. To that end, he worked diligently to perfect his mind and body. Never satisfied with simply resting on his natural talents, he read voraciously on military topics, mastered the technical elements of the military profession, and undertook a rigorous regimen of physical training to ensure that he would be prepared to shape his destiny.

Patton truly believed that leadership above all other factors won battles, and he worked diligently to maximize his innate abilities. These efforts would pay remarkable dividends for Patton later in life as he exuded confidence and calm during moments of fear and uncertainty. Indeed, many of Patton's greatest moments, such as his dramatic revitalization of Second Corps after its defeat at the Battle of Kasserine Pass in 1943, the Operation Cobra breakout in Normandy in 1944, and his relief of Bastogne five months later, came in times of crisis. When others needed inspiration, Patton was able to convince them that he could lead them to victory.

Much like his heroine, Joan of Arc, Patton used his mystic "warrior soul" to inspire courage and determination in his men when they needed it most. This article is perhaps the closest written example of Patton's leadership principles, and it provides a surprisingly profound view into the spirit of the great general.

SUCCESS IN WAR

Cavalry Journal, January 1931

War is an art and as such is not susceptible of explanation by fixed formulae. Yet from the earliest time there has been an unending effort to subject its complex and emotional structure to dissection; to enunciate rules for its waging; to make tangible its intangibility. As well strive to isolate the soul by the dissection of the cadaver as to seek the essence of war by the analysis of its records. Yet, despite the impossibility of physically detecting the soul, its existence is proven by its intangible reflection in acts and thoughts.

Above armed hosts there hovers an impalpable something which on occasion so dominates the material as to induce victory under circumstances quite inexplicable. To understand this something we should seek it in a manner analogous to our search for the soul; and so seeking we shall perchance find it in the reflexes produced by the acts of the Great Captains.

But, whither shall we turn for knowledge of their very selves? Not in the musty tomes of voluminous reports or censored recollections wherein they strove to immortalize their achievements. Nor yet in the countless histories where lesser, wormish men have sought to snare their parted ghosts.

The great warriors were too busy and often too inapt to write contemporaneously of their exploits. What they later put on paper was colored by strivings for enhanced fame, or by political conditions then confronting them. War was an ebullition of their perished past. The violent simplicity in execution which procured them success, and enthralled the world, looked pale and uninspired on paper, so they seasoned it.

The race yearns to adore. Can it adore the simple or venerate the obvious? All mythology and folklore rise in indignant protest at the thought. The sun gave light; therefore he was not hot gas or a flame, but a God or a chariot. The "ignus fatuus" deluded men of nights. It was a spirit; nothing so simple as decomposition could serve the need.

So with the soldier, to pander to self love and racial urge he attributes to his acts profound thoughts which never existed. The white hot energy of youth, which saw in obstacles but inspirations and in the enemy but the gage to battle, becomes too complacent and retrospective with age. The result of mathematical calculation and metaphysical erudition; of knowledge he never had and plans he never made.

With the efforts of the historians the case is even worse. Those who write at the time are guilty of partisanship and hero worship. While those who write later are forced to accept contemporaneous myths and to view their subject through the roseate light which distance, be it that of time or space, sheds ever to deprive us of harsh truth. In peace the scholar flourishes, in war the soldier dies; so it comes about that we view our soldiers through the eyes of scholars and attribute to them scholarly virtues.

Seeking obvious reasons for the obscure, we analyze their conduct as told by historians and assign as reasons for their success apparent, trivial things. Disregarding wholly the personality of Frederick we attribute his victories to a tactical expedient, the oblique order of battle. Impotent to comprehend the character of Rome's generals, a great historian coins the striking phrase, "At this time the Roman legionary shortened his sword and gained an empire." Our research is further muddled by the fabled heroism of all former fighters. Like wine, accounts of valor mellow with age, until Achilles dead three thousand years stands peerless.

Yet, through the murk of fact and fable rises to our view this truth. The history of war is the history of warriors; few in number, mighty in influence. Alexander, not Macedonia, conquered the world. Scipio, not Rome, destroyed Carthage. Marlborough, not the Allies, defeated France. Cromwell, not the Roundheads, dethroned Charles.

Were this true only of warriors we might well exclaim, "Behold the work of the historian!" but it is equally the case in every phase of human endeavor. Music has its myriad of musicians but only its dozen masters. So with painting, sculpture, literature, medicine, or trade. Many are called, but few are chosen.

Nor can we concur wholly with the alluring stories in the advertising sections of our magazines which point the golden path of success to all and sundry who will follow some particular phase of home education they happen to advocate. "Knowledge is power," but only to a degree. Its possession per se will raise a man to mediocrity but not to distinction. In our opinion, indeed, the instruction obtained from such courses is of less moment to future success than is the ambition which prompted the study.

In considering these matters, we should remember that while there is much similarity there is also a vast difference between the successful soldier and the successful man in other professions. Success due to knowledge and personality is the measure of ability in each case; but to all save the soldier it has vital significance only to the individual and to a limited number of his associates. With the soldier, success or failure means infinitely more, as it must of necessity be measured not in terms of personal honor or affluence, but in the life, happiness, and honor of his men; his country. Hence, the search for that elusive secret of military success, soul, genius, personality; call it what you will; is of vital interest to us all.

As has been shown, history and biography are of but limited assistance and the situation is still further complicated by other circumstances which we shall now discuss. First, we must get an harmonious arrangement between two diametrically opposed views; namely, that there is "nothing new under the sun" and that there is "nothing old."

Referring to the first assumption, that of immutability, we refer to the tendency to consider the most recent past war as the last word, the sealed pattern of all future contests. For this theory we of the military profession are largely to blame. First, we realize, none better, that in the last war it was necessary to make many improvisations and to ply our trade with ill assorted tools. We then read our books and note with a thrill of regret that in the war next preceding our own experience, "Things ran with the precision of a well oiled machine," for so the mellowing influence of time has made it appear to our authors.

In our efforts to provide for the avoidance, in future, of the mistakes which we personally have encountered, and to insure to ourselves or to our successors the same mathematical ease of operation of which we have read, we proceed to enunciate rules. In order to enunciate anything we must have a premise. The most obvious is the last war. Further, the impressions we gained there were the most vivid we have ever experienced; burned on the tablets of our memories by the blistering flash of exploding shell, etched on our souls by the incisive patter of machine gun bullets, our own experiences become the foundation of our thoughts and, all unconscious of personal bias, we base our conceptions of the future on our experience of the past.

Beyond question, personal knowledge is a fine thing; but unfortunately it is too intimate. When, for example, we recall a railroad accident, the picture that most vividly presents itself to us is the severed blue-gray hand of some child victim; not the misread signals which precipitated the tragedy. So with war experiences. The choking gas that strangled us sticks in our memory to the more or less complete exclusion of the important fact that it was the roads and consequent abundant mechanical transportation peculiar to western Europe which permitted the accumulation of enough gas shells to do the strangling.

Even when no personal experience exists, we are bound to be influenced by the most recent experience of others. Because in the Boer War the bayonet found no employment, we all but abandoned it, only to seize it again when the Russo-Japanese conflict re-demonstrated its value.

Going back further, we might point to countless other instances of similar nature, as witness to the recurrent use and disuse of infantry and cavalry as the dominant arms according to the most recent "lesson" derived from the last war based invariably on special conditions, in no way bound to recur, yet always presumed as immutable.

So much for the conservatives; now for the optimists; the "nothing old" gentry. These are of several species, but first in order of importance come the specialists.

Due either to superabundant egotism and uncontrolled enthusiasm, or else to limited powers of observation of the activities of other arms, these people advocate in the most fluent and uncompromising manner the vast future potentialities of their own weapon. In the next war, so they say, all the enemy will be crushed, gassed, bombed, or otherwise speedily exterminated, depending for the method of his death upon the arm to which the person declaiming belongs. Their spectacular claims attract public attention. The appeal of their statements is further strengthened because they deal invariably in mechanical devices which intrigue the simple imagination, and because the novelty of their schemes and assertions has a strong news interest which insures their notice by the press. Earlier examples of this newspaper tendency to exploit the bizarre is instanced in the opening accounts of the Civil War where "masked batteries" and "black horse cavalry" seemed to infest the whole face of nature.

Both the standpatters and the progressives have reason of sorts, and as we have pointed out, we must seek to harmonize the divergent tendencies.

A British writer has said, "The characteristic of war is its constant change of characteristic," but as is ever the case with aphorisms his remark needs explanation. There is an incessant change of means, to attain the inevitable end, constantly going on; but we must take care not to let these inevitable sundry means, past or predicted, attain undue eminence in the perspective of our minds. Since the beginning there has been an unending cycle of them and for each its advocates have claimed adoption as the sole means of successful war. Yet, the records of all time show that the unchanging ends have been, are, and probably ever will be, the securing of predominating force, of the right sort, at the right place, at the right time.

In seeking a premise for the enunciation of rules for the employment of this predominating force, we must cull from the past of our experience or reading the more permanent characteristics, select our weapons and assign to them that importance which reason and the analogy of experience indicate that they will attain. Bearing in mind these considerations and the definition of predominant force, we shall resume our search for success in war.

No matter what the situation as to clarity of his mental perspective, the conscientious soldier approaches the solution of his problem more or less befuddled by phantoms of the past, and deluded by unfounded or unproved hopes for the future. So handicapped, he assumes the unwonted and labored posture of a student, and plans for perfection, so that when the next war comes that part of the machine for which he may be responsible shall instantly begin to function with a purr of perfect preparation.

In this scholarly avocation, soldiers of all important nations use at the present time what purports to be the best mode of instruction; the applicatory method. The characteristics of some concrete problem are first studied in the abstract and then tested by applying them, with assumed forces and situations, in solving analogous problems either on the terrain or on a map representation of it. This method not only familiarizes the student with all the tools and technicalities of his trade, but also develops the aptitude for reaching decisions and the self assurance derived from demonstrated achievement.

But, as always, there is a fly in the amber. High academic performance demands infinite intimate knowledge of details, and the qualities requisite to such attainments often inhibit bodies lacking in personality. Also, the striving for such knowledge often engenders the fallacious notion that capacity depends upon the power to acquire such details rather than upon the ability to apply them. Obsessed with this thought, students plunge in deeper and ever deeper, their exertions but enmeshing them the more until, like mired mastodons, they perish in a morass of knowledge where they first browsed for sustenance.

When the prying spade of the unbiased investigator has removed the muck of official reports of the World War, the skeletons of many such military mammoths will be discovered. Amid their mighty remains will lurk elusive the secret of German failure. Beyond question no soldier ever sought more diligently than the German for prewar perfection. They built and tested and adjusted their mighty machine and became so engrossed in its visible perfection, in the accuracy of its bearings and the compression of its cylinders, that they neglected the battery. When the moment came, their masterpiece proved inefficient through the lack of divine afflatus, the soul of a leader. Truly in war, "Men are nothing, a man is everything."

Here we must deny that anything in our remarks is intended to imply belief in the existence of spontaneous untutored inspiration. With the single exception of the divinely inspired Joan of Arc, no such phenomenon has ever existed, and as we shall show, she was less of an exception than a coincidence. We require and must demand all possible thoughtful preparation and studious effort, so that in war our officers may be equal to their mighty trust; the safety of our country. Our purpose is not to discourage such preparation, but simply to call attention to certain defects in its pursuit. To direct it not towards the glorification of the means, study; but to the end, victory.

In acquiring erudition we must live on, not in, our studies. We must guard against becoming so engrossed in the specific nature of the roots and bark of the trees of knowledge as to miss the meaning and grandeur of the forests they compose. Our means of studying war have increased as much as have our tools for waging it, but it is an open question whether this increase in means has not perhaps obscured or obliterated one essential detail; namely, the necessity for personal leadership.

Hannibal, Caesar, Heraclius, Charlemagne, Richard, Gustavus, Turenne, Frederick, Napoleon, Grant, Lee, Hindenburg, Allenby, Foch, and Pershing were deeply imbued with the whole knowledge of war as practiced at their several epochs. But so were many of their defeated opponents; for as has been pointed out, the success in war lies not wholly in knowledge. It lurks invisible in that vitalizing spark, intangible, yet as evident as the lightning—The Warrior Soul.

There is no better illustration of the potency of this vitalizing element than is portrayed in the story of the "Maid of Orleans." For more than ninety years prior to her advent, the armies of France had suffered almost continuous defeat at the hands of their British opponents. The reason for this state of things lay not in the inferiority of French valor, but in the reappearance of the foot soldier armed with the missile weapon (the long bow) as the temporary dominating influence on the battlefield. As a result of the recurrence of this tactical condition, France suffered almost continuous defeats, with the result that her people lost confidence, and developed an inferiority complex. Then came Joan, whose flaming faith in her heaven sent mission rekindled the national spirit. Yet, great as were her powers, it is idle to suppose that, all unschooled in war as she was, she could have directed unaided the energy she produced. Like the fire beneath the boiler, she produced the steam; and ready to her hand she found competent machinery for its utilization in the shape of those veteran soldiers, Dunois, La Hire, and

Saint Railles. The happy coincidence of her ignorant enthusiasm and their uninspired intelligence produced the phenomenal series of victories which freed France.

We now shall seek to evaluate and place in their just ratio the three essentials to victory; Inspiration, Knowledge, and Force (mass).

Napoleon won many battles with numbers inferior to the enemy; he never lost a battle when he was numerically superior. In other words, even his transcendent ability was not equal, on every occasion, to the task of counterbalancing numerical inferiority. When he was confronted with the admittedly incapable Austrian generals of 1796 he destroyed armies; while later, particularly after 1805, his victories were far less overwhelming. So with Caesar. Against the Nervae, he was a consuming flame; against Romans, a successful commander. Grant in the wilderness was as nothing compared to Grant at Donaldson or before Vicksburg. Here we have three soldiers of the highest type, both mentally and spiritually. By way of contrast we may note how the learned but uninspired Prussians of 1870 triumphed over the poorly led French, while in 1914 their equally learned and uninspired descendants were far less successful in the face of better opposition.

We may therefore postulate that no one element (soul, knowledge, or mass) is dominant; that a combination of any two of these factors give a strong presumption of success over an adversary who relies on one alone, while the three combined are practically invincible against combinations of any other two. Comparing our own resources as to mass with those of any possible opponent or group of opponents, we strike at least a balance. The demonstrated ability of our trained leaders in past wars shows that so far as education is concerned, our officers have no superiors and few equals. This being so, victory will fly to or desert our standards in exact proportion to the presence or absence, in our leaders, of the third attribute.

War is conflict; fighting is an elemental exposition of the age old effort to survive. It is the cold glitter of the attacker's eye, not the point of the questing bayonet, that breaks the line. It is the fierce determination of the driver to close with the enemy, not the mechanical perfection of the tank, what conquers the trench. It is the cataclysmic ecstasy of conflict in the flier, not the perfection of his machine gun, which drops the enemy in flaming ruin. Yet, volumes are devoted to armament; pages to inspiration.

Since the necessary limitations of map problems inhibit the student from considering the effects of hunger, emotion, personality, fatigue, leadership, and many other imponderable yet vital factors, he first neglects and then forgets them. Obsessed with admiration for the intelligence which history has ascribed to past leaders, he forgets the inseparable connection between plans, the flower of the intellect, and execution, the fruit of the soul. Hooker's plan at Chancellorsville was masterly, its execution cost him the battle. The converse was true at Marengo. The historian, through lack of experience and consequent appreciation of the inspirational qualities of generals, fails to stress them but he does emphasize their mental gifts, which, since he shares, he values. The student blindly follows, and hugging the notion of mentality, pictures armies of insensate pawns moving with the precision of machines and the rapidity of light, guided in their intricate and resistless evolutions over the battlefield by the cold effulgence of his emotionless cerebrations as transmitted to them by wire and radio through the inspiring medium of coded messages. He further assumes that superhuman intelligence will translate those somber sentences into words of fire which will electrify his chessmen into frenzied heroes who, heedless of danger, will dauntlessly translate the stillborn infants of his brain into deeds.

Was it so that Caesar rallied the Twelfth Legion? Could the trackless ether have conveyed to his soldiers the inspiration that Napoleon imparted by his ubiquitous presence when before Rivoli he rode five horses to death, "To see everything for himself"? Staff systems and mechanical communications are valuable, but not above and beyond them must be the commander; not as a disembodied brain linked to his men by lines of wire and waves of ether, but as a living presence, an all pervading visible personality. The unleavened bread of knowledge will sustain life, but it is dull fare unless leavened by the yeast of personality. Could seamanship and shooting have made the Bon Homme Richard prevail over the Serapis or have destroyed the French fleet in Abukar Bay, had Paul Jones and Horatio Nelson been other than they were? What intellectual ghost replete with stratagem could have inspired men as did these two, who in themselves have epitomized not only knowledge of war, but the spirit of battle? In defining the changeless characteristics of war we mentioned force, place, and time. In our calendar of warriors, Napoleon Bonaparte and Stonewall Jackson stand preeminent in their use of the last of these; time. Of the first, his soldiers boasted, "He wins battles more with our legs than with our bayonets," while Jackson's men proudly called themselves "Old Jack's foot cavalry."

Shrewd critics have assigned military success to all manner of things; tactics, shape of frontiers, speed, happily placed rivers, mountains or woods, intellectual ability, or the use of artillery. All in a measure true, but none vital. The secret lies in the inspiring spirit which lifted weary, footsore men out of themselves and made them march, forgetful of agony, as did Messena's division after Rivoli and Jackson's at Winchester. No words ever imagined could have produced such prodigies of endurance as did the sight of the boy general, ill, perched on his sweating horse, or of the stern puritan plodding ever before them on Little Sorrel. The ability to produce endurance is but an instance of that same martial soul which arouses in its followers that resistless emotion defined as "élan" the will to victory. However defined, it is akin to that almost cataleptic burst of physical and mental exuberance shown by the athlete when he breaks a record or plunges through the tacklers, and by the author or artist in the creation of a masterpiece. The difference is that in the athlete or the artist the ebullition is auto-stimulated, while with an army it is the result of external impetus; leadership.

In considering war we must avoid that adoration of the material as exemplified by scientists who deny the existence of anything they cannot cut or weigh. In war tomorrow we shall be dealing with men subject to the same emotions as were the soldiers of Alexander; with men but little changed for better or worse from the starving, shoeless Frenchmen of the Italian Campaign; with men similar, save in their arms, to those whom the inspiring powers of a Greek or a Corsican changed at a breath to bands of heroes, all enduring and all capable.

No! History as written and read does not divulge the source of leadership. Hence, its study often induces us to forget its potency. As a mirror shows us not ourselves, but our reflection, so it is with the soul and with leadership; we know them but by the acts they inspire or the results they achieve. Like begets like; in the armies of the great we seek the reflection of themselves and we find Self Confidence, Enthusiasm, Abnegation of Self, Loyalty, and Courage.

Resolution, no matter how adamant, mated to knowledge, no matter how infinite, never begat such a progeny. Such offspring arises only from blood lines as elemental as themselves. The leader must be incarnate of them.

The suggestion of Nicodemus as to rebirth (John III, 3–6) is not the only means of producing such a leader. There are certainly born leaders, but the soldier may also overcome his natal defects by unremitting effort and practice. Self confidence of the right sort as differentiated from bumptious presumption based on ignorance, is the result of proven ability, the sense of conscious achievement. Its existence presupposes enthusiasm, for without this quality no one could endure the travail of acquiring self confidence. The enthusiasm which permits the toil and promises the achievement is simply an all absorbing preoccupation in the profession elected. Endurance, too, is linked with self confidence. Mentally it is the ability to subvert the means to the end, to hitch the wagon to a star and to attain it. Physically it presupposes sufficient enthusiasm to force on nature, no matter how reluctant, the obligation of constant bodily fitness through exercise. The expanding waistline means the contracting heart line; witness Napoleon at and after Jena. Abnegation of self seems perhaps incongruous when applied to such selfish persons as Frederick or Napoleon, but this is not the case. Self can be subordinated to self. The Corsican, leading his grenadiers at Lodi, subordinated the life of Bonaparte to the glory of Napoleon. Loyalty is frequently only considered as faithfulness from the bottom up. It has another and equally important application, that is from the top down. One of the most frequently noted characteristics of the great who remained great is unforgetfulness of and loyalty to their subordinates. It is this characteristic which binds with hoops of iron their juniors to them. A man who is truly and unselfishly loyal to his superiors is of necessity so to his juniors, and they to him.

Courage, moral and physical, is almost a synonym of all the foregoing traits. It fosters the resolution to combat and cherishes the ability to assume responsibility be it for successes or failures. No Bayard ever showed more of it than did Lee after Gettysburg.

But, as with the biblical candle, these traits are of no military value if concealed. A man of diffident manner will never inspire confidence. A cold reserve cannot begat enthusiasm and so with the others there must be an outward and visible sign of the inward and spiritual grace.

It then appears that the leader must be an actor, and such is the fact. But with him, as with his bewigged compeer, he is unconvincing unless he lives his part.

Can men then acquire and demonstrate these characteristics? The answer is, they have; they can. For, "As a man thinketh, so is he." The fixed determination to acquire the warrior soul, and having acquired it, the determination to either conquer or perish with honor, is the secret of success in war.

FIVE

Anticipating the Next War

Well, hat are the lessons of history, and what do they
tell us about future wars? Patton was fascinated by
these questions throughout his professional life
and attempted to tackle them in his 1932 Army War College
thesis, "The Probable Characteristics of the Next War and
the Organization, Tactics, and Equipment Necessary to Meet
Them." In working on his thesis, Patton was given very broad
latitude and was allowed to write at a leisurely pace on a
"topic of interest to the War Department." This process may
have lacked the structure of the current military education
system, but it produced a series of excellent studies during the
interwar period, by Patton and others, which correctly antici-
pated future contingencies in Germany and Japan.[1]

Rather than being buried in a War Department file cabi-
net, Patton's thesis was commended by the War College commandant as a "work of exceptional merit." By virtue of this
high praise and official blessing, the paper circulated widely
in top American defense circles and helped disseminate Patton's views to a broader military audience. In part because of
his thesis, Patton graduated with honors from the War College in 1932, which helped solidify his reputation as one
of the Army's top majors and among its next generation of
thinkers.[2]

The thesis showcases two of Patton's greatest skills,
his deep study and understanding of military affairs and his

seemingly clairvoyant ability to anticipate future contingencies on the basis of this knowledge. Given the volatile changes and extreme uncertainty in military affairs during the interwar period, this document is truly astonishing.[3] In this relatively short study, Patton correctly predicts the major developments in combined warfare that occurred during the interwar period and creates a clever strategy to use America's limited interwar budget to exploit these new opportunities.

Patton argues that large mass armies often result in high casualties and protracted wars. Rather than using limited U.S. resources to build a mass army, therefore, Patton advocates for the creation of a smaller professional force in order to restore movement, and with it decisive victory, to the battlefield.

Based on his assumptions, Patton makes many prescient forecasts in his thesis, one of the most remarkable being that the Treaty of Versailles would not be able to curtail the military ambitions of Germany. Prior to the rise of Hitler, few military men imagined that the small professional German army would be useful for anything beyond border defense and internal security. Patton, however, notes the innovations and reforms General Hans von Seeckt made during the 1920s as something more than "simply making the best of a bad bargain" and correctly predicts the danger inherent in a small but excellent cadre of military experts.

For Patton, a small but professional German military was far more dangerous than a large conscript army because the smaller force would be better suited to the complex and rapid character of future war and thus more likely to win a quick and decisive victory.[4] But this anticipation of the rise of maneuver warfare and the ability to achieve rapid victory was very much out of step with the considered military opinion of the period. Despite the recent deadlock of World War I, Patton was confident that new weapons and tactics combined with professional armies could restore movement and decisiveness to warfare. For a decade and a half, Patton had believed that tanks had the ability to overcome defensive and natural obstacles and serve as the arm of decision in future conflicts. Although he bemoaned the seeming indifference that American strategists had for armor during this period, he correctly believed that other powers were making great strides in tank doctrine and training.

Expanding on themes he had developed in his previous works, Patton used his study to again argue for the importance of human factors such as training, courage, and leadership. Despite his insistence on the importance of new technology and doctrine, Patton believed that the majority of American manpower should be allocated to the infantry. In addition to saving money during a period of lean budgets, Patton believed that a more conventional force structure would avoid many of the technical glitches associated with the incorporation of new equipment. He noted that since the United States was already the most automobile-friendly country in the world, it would have little problem producing vehicles or training professional soldiers for the mechanized battlefield. In addition, he argued for the constant testing and acquisition of small numbers of new tank models so the Army would have the benefit of new technology without a large number of legacy platforms (a problem that would particularly hamper the French tank force during the interwar period).

Overall, Patton's thesis is a remarkable combination of historical analysis and a prescient grasp of the future strategic environment. In places, however, Patton makes selective use of history to advocate for the superiority of professional armies. For example, he claims that the Roman mass armies were inferior to the armies of both Pyrrhus and Hannibal, focusing on these generals' tactical successes. Patton seemingly ignores the fact that Rome ultimately triumphed because of the greater resiliency of their mass armies compared to the more technically proficient but less durable professional forces of its foes. In fact, Pyrrhus is best known today as the general who won a battle but at an unacceptable cost to his small force, thus coining the term "Pyrrhic victory."

Despite these minor inconsistencies, Patton's thesis is a document that provides an invaluable insight into Patton as a man of thought. The ability to anticipate the enemy's next move would become a Patton trademark. In addition to the claims of this 1932 study, Patton's intelligence reports and other published works predicted a series of events ranging from the surprise Japanese attack on Pearl Harbor to the desperate German attack through the Ardennes in 1944. Patton appeared to have a sixth sense for war because he thought multiple steps ahead rather than just responding to the enemy's moves. This gave him the initiative in almost every battle and allowed him to move faster and more decisively than his opponents. Indeed, much of the general's success can be directly linked to his rare combination of intellectual talent and strategic foresight. As such, this thesis should be studied as both an excellent example of professional military education and an insight into Patton's ability to translate his studies into a clear vision of the future.

THE PROBABLE CHARACTERISTICS OF THE NEXT WAR AND THE ORGANIZATION, TACTICS, AND EQUIPMENT NECESSARY TO MEET THEM

1932

I. INTRODUCTION

"All experience hath shown, that mankind are more disposed to suffer, while evils are sufferable, than to right themselves by abolishing the forms to which they are accustomed."

In these flowing words the brilliant author of the "Declaration of Independence" gave expression to the fact that the human mind prefers to remember rather than to think; to endure rather than to adventure.

Due to this habit we tend to have an excessive admiration for the past, and frequently carry our veneration to the point of believing that it also depicts the future.

The widespread opinion that the World War, waged as it was, in complete accord with the principle of "The Nation in Arms," is a new development and the sealed pattern for future wars, is a case in point. As a matter of fact, the principle of the "Nation in Arms," is older than history.

During the forty-four hundred years that separate the Syrian invasion of Egypt from the German invasion of France, there have been countless wars waged on the mass system; and a practically equal number conducted, on the diametrically opposed principle, inherent in the use of professional armies.

Now, while there is a strong school of military thought which holds that all historical study prior to 1870 is futile, the apparently inexorable recurrence of the cycles of history is so impressive as to merit investigation. Without perspective, a painting is valueless; so it is with things military.

Unquestionably, it is foolish to copy ancient tactics, but we should familiarize ourselves with the causes which impelled their adoption, because in the four thousand odd years of recorded history man has changed but little. Save for appearances, the hoplite and the rifleman are one, and the emotions and consequent reactions which affected one affect the other.

This being so, it behooves us to pause for a moment and classify old wars according to the means used in their waging.

II. TYPES OF ARMIES USED IN FORMER WARS

GENERAL

In order to simplify this investigation, nomenclature has been standardized as follows:

Armies composed of men maintained, equipped, and trained over a period of years, for the sole purpose of war, are called "professionals."

Armies composed of men, however maintained, equipped, and trained, who make war a secondary consideration, are called "mass armies."

In applying these general terms it must be remembered that they are relative and should be interpreted in consonance with the cultural period in which the forces operated.

Clearly, the amount of time, training, and money necessary to produce an Egyptian Bowman of 1500 B.C., a Roman Legionnary of 45 B.C., a French Grenadier of 1796, or a British Regular of 1914, varied in an ascending scale. Yet, they are one in that a group of any of them was superior to an equally numerous group of contemporary amateurs.

The basic difference between professional and mass armies is the difference between quality and quantity; these attributes cannot be combined.

OLD WARS CLASSIFIED AS TO TYPES OF ARMY USED

2500 B.C. Mass Army. One of the earliest wars of which there is authentic record occurred between Egypt and Syria during the Sixth Dynasty. A force of unknown character from Syria attacked Egypt. To meet this, the Egyptians raised a levee in mass, calling on each province "None" to furnish its quota. (Erman.)

Lesson. A short campaign of home defense can be conducted with a mass army. At this period tools and weapons were simple and often identical.

2000 B.C. Professional Army. Under the Middle Empire, Egypt sought to maintain peace by isolation. To attain this end, fortification was invented and highly perfected. Two great forts, one at Assuan and the other at Pelusium, were built and manned by professional soldiers. (Breasted.)

Lesson. Complicated equipment demanded professionals.

1500 B.C. Mass Army. At this time Assyria came to the forefront. All males of military age were liable to serve in the army. (Spaulding.)

Lesson. Wars were local; weapons simple.

1400 B.C. Professional Army. Thothmes III invaded Syria with an army of 15,000 men. This force marched from the present site of the Suez Canal to Mount Carmel, 250 miles in twenty-two days. From this fact and from the account of the battle fought at Armageddon, it is believed that the force was composed of professionals. (Petrie.)

Lesson. Distant wars and hard campaigning need quality rather than quantity.

1000 B.C. Mass Army. At this time the Greek States had wholly mass armies, all males being required to serve. The chiefs, and some of the lesser chiefs, being better armed and more practiced in war, were much more susceptible of being classed as "professionals."

Lesson. Good weapons were costly and, hence, limited to the small professional class.

722 B.C. Professional Army. As we have seen, Assyria started with a mass army. Under Sargon II, her army consisted of 50,000 professionals. In peacetime this force was used to garrison provinces and protect the border. For wars it was augmented by an additional mass army of 150,000 men. (Breasted.)

Lesson. Distant operations need professionals.

650 B.C. Professional Army. Greek mercenaries were used in Egypt. (Herodotus.)

Lesson. The Egyptians had money, but had no military ability; so they hired it.

600 B.C. Mass Army. By this time all the Greek States had a well defined method of universal service, citizens and non-citizens being subject to call. Note. Non-citizens were not allowed heavy armor. (Westermann.)

Lesson. Wars were local and of short duration. The wealthy Greek citizens owned their equipment.

546 B.C. Professional Army. Under Cyrus and Cambyses, the Persian Army was professional. In very large wars, either of conquest or of defense, its numbers were greatly augmented by local militia. (Herodotus.)

Lesson. Distant operations and continuous wars demand professionals.

460 B.C. Professional Army. Due to constant wars, all the Greek States began to employ mercenaries. (Spaulding.)

Lesson. Civilians could not attend to business and at the same time remain constantly in the army. At this time arms and equipment became the property of the State. In order to keep the mercenaries from wasting them, the rules for discipline became much more rigid.

480 B.C. Professional and Mass Armies. At Thermopylae the Greek Army of 7,000 consisted of a trained national levee. However, the Spartan contingent of less than a thousand probably should be called "professional," due to the fact that their whole useful military life was spent in the practice of arms. In the Persian Army of 100,000 the Guard Corps of 10,000, called the "Immortals" because they were always kept up to strength, were the only professionals. (Cressy.)

370 B.C. Professional Army. Philip of Macedon hired Greek mercenaries, not only to form part of his army but also to act as models for the rest which was composed of his subjects. Alexander the Great used this army, which was purely professional. With it he conquered the known world and in every battle defeated forces composed of mass armies which greatly outnumbered him. (Herodotus.)

Lesson. For wars of conquest and distant campaigns professional armies are necessary.

280 B.C. Professional and Mass Armies. The Romans, with a good mass army, were defeated by Pyrrhus with a professional army. (Spaulding.)

Lesson. Quality is superior to quantity.

218 B.C. Professional and Mass Armies. Hannibal, with a mercenary army composed of hired men from many nations, repeatedly beat Roman mass armies largely superior to him in numbers. After Cannae (216 B.C. Mass Army), Scipio organized a new mass army. He placed so much emphasis on drill, organization, and equipment that, at Zama, he defeated Hannibal's professional army. Note. The bad behavior of the Carthaginian Cavalry, which, in this case, was not professional, was in large measure responsible for the Carthaginian defeat. (Polybius.)

Lesson. Quality triumphed over quantity. For an overseas operation the semiprofessional army was needed. This Scipio produced. Due to the fact that it was homogeneous and national, it proved superior to the veteran mercenaries of Hannibal.

105 B.C. Professional Army. The Cimbric and Teutones, a perfect exemplification of the "Nation in Arms," destroyed two Roman armies at Arausio (Orange) on the Rhone, 105 B.C. Marius then organized a new Roman Army, enlisted for sixteen years, and with it, in 102 B.C., annihilated the Teutones, near Aix. (Oman.)

Lesson. Emotional enthusiasm can, on occasion, defeat discipline. Since the Romans could not match the Teuton masses, and since their militia army could not stand the Teutones' ardor, it was necessary for them to find a new means. This they attained in the rigid discipline of the sixteen year enlistment. Another cause leading them to this was that the constant disbanding of armies at the end of each war failed to utilize the military training of the veterans. Note. The system of numbered legions in the Roman armies started at this time. (Oman.)

59 to 44 B.C. Professional and Mass Armies. Caesar, utilizing the rapid marching and high battle mobility of his professional armies, defeated many mass armies, all of which invariably outnumbered him. In the civil wars his victories were much less striking, since here both sides used professionals. (Caesar's commentaries.)

Lesson. Quality superior to quantity. Similarity of type of army or of tactics has always produced indecisive results.

29 B.C. to 380 A.D. Professional Army. The professional armies of Rome engaged in constant wars; during the whole of this period with vastly superior mass armies. They were almost invariably successful. However, in 251 A.D., the Goths defeated a Roman army, under Decius, at Trebonii. (Oman.)

Lesson. The Roman infantry armies were more mobile than most of their foes. The mounted Goths were more mobile than the Romans. In this latter case their mobility and enthusiasm more than made up for their lack of training.

378 A.D. Professional and Mass Armies. At the battle of Adrianople the last Roman army of the old type was utterly defeated by the Gothic mass army. From this date onward, for 1,000 years, Cavalry replaced Infantry as the dominant arm of battle. (Oman.)

Lesson. Again mobility and enthusiasm more than compensated for lack of training. Also, the Roman army was decadent.

530 A.D. Professional Army. By this time, the army of the Eastern Empire was wholly professional and consisted of mounted bowmen, using both fire and shock. Infantry elements, when needed for defensive operations or sieges, were raised on the mass system. With such a professional army, Belisarius defeated, at Daras, a mass Persian army double his number, and five years later with an army of 15,000 professionals conquered all of northern Africa. His opponents were mass armies. (Oman.)

Lesson. Superior mobility and discipline.

Europe 500 to 800 A.D. Mass Armies. In Europe during this period the art of war sunk to a very low ebb. All fights were local and were conducted by small levees. Some few of the Counts and their clients may claim professionalism due to their frequent practice in wars. (Oman.)

756 to 850 A.D. Semi-Professional Army—Charlemagne. "Charles the Great undertook offensive wars on a much larger scale than Pepin and Charles Martel. His armies went far afield and the regions he subdued were so broad, that the old Frankish levee in mass would have been far too slow and clumsy a weapon for him. To keep this mighty empire in obedience a more quickly moving force was required. Hence, Charles did his best to increase the number of his horse soldiers." (Oman.)

To insure this result he proceeded as follows: In 779 he passed a law prohibiting the exportation of armor. In 803 he reduced the size of his army by arranging all citizens into groups according to wealth, varying from five to three men each, and requiring that each group supply one man armed, mounted, and equipped. By the laws of 805 and 807 he still further stressed quality over quantity by increasing the size of the groups and demanding battle equipment from the man sent. In 811 he promulgated a code of military punishments. In 813 he passed a law specifying not only the number and equipment of the soldiers, but also the kind of transportation, engineer stores, quartermaster, and ordnance property that should accompany each unit. To hold the ground that he had conquered he created a system of fortified camps connected by roads; they were called "Burgs." He garrisoned these burgs with military set- tlers, whom he provided with farms and wives. (Charlemagne; *First of the Moderns.* Russell.)

Lesson. This attempt at professionalism was highly successful.

659 to 1071 A.D. Semi-Professional Army—Eastern Empire. At the close of the Saracen war in 659, Constantine the Great organized his empire into Corps Areas, maps of which are still available. These Corps Areas were called "Themes." Each "Theme" was garrisoned by a corps of from eight to twelve thousand professionals. After deducting fortress troops and border guards each "Theme" could produce a field force of 6,000 mounted men. These forces were organized into the most minute detail; from the squad of ten men up, including the division. They had medical and supply units and arsenals. The Emperor Maurice, while still a general under Constantine, wrote a manual for general officers called "Strategicon." Not only are all forms of tactics covered in this manual, but different types of strategy applicable against the several enemies of the empire are specified in great detail.

Note. The chapter in this book with reference to the examination of prisoners of war is almost identical in words with the regulations used by us in France in 1917.

This manual was rewritten and brought up to date in 900 by Leo the Wise. Professional armies, organized in accordance with these regula- tions, maintained the integrity of the empire until 1071, at which date the army was badly beaten by a vastly superior mass army of Turks at Manzikert. After this defeat the "Theme" system fell into disuse and the army degenerated into a mass of mercenary bands, using their own tac- tics and equipment. (Oman and "The Strategicon.")

Lesson. There is nothing new.

800 to 850 A.D. Mass and Professional Armies—Vikings. The Vikings, who initially were volunteer robbers, became professionals through experience. They stole horses and so gained such mobility as to be perfectly immune from the mass levies sent against them. About 900, feudalism began to evolve as an antidote to the Vikings. At first some leader would guarantee the protection of a part of the country provided that the inhabitants would pay him. With this money he hired, armed, and equipped a small body of professionals. Such forces defeated the Vikings because they not only marched but also fought on horseback, whereas the Vikings dismounted to fight and had no missile weapons. By 1000 A.D., these feudal lords had developed to such a state of military efficiency that they could defeat any number of peasants. Froissart states that sixteen of them defeated 2,000 peasants in one afternoon.

Lesson. Superior mobility of professionals.

Europe 1097 to 1271 A.D.—Mass Armies. During this period feudalism was triumphant. Most of the wars were local. The armies were of the mass type, stiffened by the professional followers of the feudal lords. The Crusaders were purely mass armies. However, the troops who fought under Baldwin of Jerusalem were professionals who elected to stay. The striking success which Baldwin had with his tiny forces against vastly superior numbers of Saracens and Egyptians is eloquent of the value of professionals. (Chronicles of the Crusades.)

In 1173, Henry II of England, then engaged in war with France, found that feudal levees were expensive and inefficient. He therefore desired to hire mercenaries. The feudal levee was obliged, by law, to serve forty days. Neither the peasants nor the noblemen were particularly anxious to cross the seas to France. Henry utilized this fact by calling out the National Levee and then exempting all men, who did not care to serve, at the rate of 2 shillings and 8 pence apiece. With this money, which was called "Soutage" (shield money), he hired his mercenaries. In addition to such mercenaries, every castle had a very small professional garrison.

Note. The great castle of Chateau Gaillard was defended for many months by a garrison of 300 English professionals against an army of 10,000 Frenchmen.

It is probable that the success gained by the English at the beginning of the Hundred Years War arose largely from the fact that since they were on an overseas expedition their troops were largely professionals, whereas the French being at home used mass armies. Towards the end both sides used professional armies and the results became very indecisive.

The English defeated the French at Crecy in 1346 A.D. They accomplished this by dismounting their knights and occupying a defensive position with them and with archers. The French charged into this and were shot down without being able to close. This put an end, for a time, to the dominance of Cavalry.

Lesson. Overseas operations demand professionals.

1315 to 1515 A.D. Professional Army. Swiss mercenary infantry, using halberds and pikes, became the main reliance of all armies. Their eventual disappearance was due to lack of discipline and to firearms.

Eastern Europe 1230 to 1350 A.D. Mobile Mass Army— Genghis Khan. By the use of higher mobility the Mongols overran many weak nations. However, the strong Sultans of Egypt defeated them and they were finally turned back by the walled towns and forts along the Oder and Drave. Their constant experience in war probably justifies their being classed as "Semi-Professionals." (Lamb.)

Lesson. Mobility and enthusiasm are a powerful combination.

1350 A.D. Professional Army. The Turks started the use of Janissaries. This was a form of professional guard corps which never exceeded 10,000 men. (Oman.)

1446 A.D. Professional Army. First Standing Army. Charles VII of France raised twenty Compaignes d'Ordonnance. These were mounted units, each consisting of 200 armored lancers, 200 armored and mounted archers or crossbowmen, 200 unarmored archers and horseholders. To provide an infantry to back this force, Louis XI organized a so-called "Francais [*sic*] Archers." This force was a paid, but not drilled militia.

Lesson. Complicated weapons and tactics made the use of professionals necessary.

1469 A.D. Mass Army. During the "War of the Roses," the English were at home and, despite their century of experience with professionals, they immediately reverted to the use of levees.

1494 A.D. Professional Army. The French Army which invaded Italy consisted of 25,000 professional cavalry, 12,000 Swiss infantry, and 30,000 militia infantry. They were opposed by mercenaries.

1469 A.D. Professional Army. Spain organized a standing army.

1566 A.D. Professional and Mass Armies. Wars in the Netherlands. The Spanish professional army was opposed by a national militia and mercenaries. In 1585, Maurice of Nassau tried with success to raise the tone of his militia army by reducing the size of the units and making much more rigid the discipline and drill. (rise and fall of the Dutch Republic.)

1618 to 1648 A.D. Professional Army. The "Thirty Years War" was fought with relatively small armies of mercenaries on both sides.

Lesson. Similarity of organization, tactics, and equipment produced a long indecisive war.

1642 A.D. Professional and Mass Armies. The Civil War in England began with the royal forces consisting of untrained volunteers and a few mercenaries, and a parliamentary force of organized but untrained militia.

1645 A.D. Professional Army. In 1645, Cromwell commenced the organization of the New Model Army; a Professional Force.

Lesson. Triumph of professionals.

1700 A.D. Professional Army. In the wars of the Spanish and Austrian Secessions, both sides used professional armies. However, the ravages due to disease and the practice of partly demobilizing every winter prevented these forces from arriving at any high state of drill or efficiency.

Lesson. Similarity of organization, tactics, and equipment produces long and indecisive wars.

1740 A.D. Professional Army. Frederick the Great had a highly trained army of 80,000 men, enlisted for life. His wars were fought with such a force against other professional armies whose training, however, was far less effective than his own. (Carlyle.)

Lesson. High efficiency, coupled with superlative leadership, equalized numerical inferiority.

1792 to 1815 A.D. Mass and Professional Armies. Marshal Foch states that Valmy is the first battle in which a "Nation in Arms," as now understood, appears.

The wars of the French Revolution and First Empire were fought by this type of army against the professional armies then in vogue. It is noteworthy that, due to long war experience and the enthusiasm of reformers, the French Armies attained very high ability which, when coupled with the genius of Napoleon, made them long invincible. It is further important to note that, while he had these efficient troops, he relied on mobility rather than numbers, particularly in his tactics. When they became extinct he had to resort to mass tactics. Of his troops of 1813 he said, "With recruits it is possible to win battles, but not campaigns."

Lesson. Mass armies imbued with enthusiasm, using new tactical methods and being superlatively led, can defeat professionals. Genius without proper tools must eventually fail.

1861 to 1865 A.D. Mass Armies. In the Civil War both sides used identical organizations and tactics.

Lesson. Identical methods produce long wars.

Up until the summer of 1863 a regular force on either side would have had decisive results. After that date both sides were professional in everything except discipline. Note. In 1864, Lee wrote a long order on the necessity for securing discipline. (Henderson.)

The initial successes of the South were largely due to the fact that superior enthusiasm (emotional urge) replaced discipline. In the North this enthusiasm was less marked, especially in the eastern armies.

1870 to 1871 A.D. Mass and Professional Armies. In this war a very efficient, numerous, and enthusiastic Mass Army, excellently led, easily defeated a numerically very inferior Professional Army, badly organized and led with most remarkable inefficiency.

In 1871, the surprising results gained by the new French Army are noteworthy. (Moltke.)

Lesson. Novelty of organization, combined with usable numerical superiority and good leadership, defeated a poor professional army. Note. If the commanders had been swapped a year before the war started, the results would possibly have been reversed.

1899 A.D. Mass and Professional Armies. The war in the Transvaal hardly fits the headings used due to the fact that in this case the mass army was numerically much inferior. Its chief value comes from the lesson it gives in the lag between new weapons and new tactics. (German Official History.)

Lesson. Great mobility in a large theater of war, combined with new weapons and methods and opposed to stupid leadership and obsolete tactics, is bound to secure results out of all proportion to the means used.

Summary of lessons as to types:

The conclusions deducible from the above summary may be tabulated as follows; Conditions tending to the use of:

Professional Armies	*Mass Armies*
1. Complicated equipment	1. Simple equipment
2. Costly equipment	2. Cheap equipment
3. Intricate & precise formations	3. Simple tactical formations
4. Necessity for mobility	4. Mobility not needed
(*Note:* With the exceptions of the opposite, professional armies have always had a higher mobility than masses)	(*Note:* In cases of Mongols, Arabs, Sythians, Aggyars, Boers, etc., this was not true)
5. Distant operations	5. Local wars (*Note:* Now out of use)
6. Necessity for rapid decision absent	6. Necessity for rapid decision absent
7. Protracted operations	7. Short, seasonal, inevitable wars
8. Supply difficult	8. Supply was easier
9. Discipline is more important than emotional inspiration	9. Emotional inspiration replaced the cohesive power of discipline

Note. With reference to items 1, 2, and 3, column 1 above, the complexity and cost of equipment are now far higher, both relatively and absolutely, than at any time in history. To meet modern conditions tactical formations are more intricate and make higher demands on the individual than ever before.

III. RECENT EVENTS AND OPINIONS AS TO TYPES

General

So much for history. Let us now examine more current events and the opinions of contemporary soldiers.

Treaty of Versailles

In the first place, we have the striking coincidence offered by the diametrically opposed effects induced by the treaties of Tilsit and Versailles.

By the former, a defeated Prussia was stripped of her professional army; she answered by the recreation of a national one. By the latter, a greater Prussia was deprived of universal service. Is it not probable that the energy which made her conscripts formidable will do the same for her professionals?

The following statements would seem to substantiate such a possibility:

On March 3, 1919, the military advisors to the peace conference headed by Marshal Foch submitted a recommendation that, for the future, Germany be limited to an army of 200,000 men, recruited either by voluntary enlistment or by conscription for a period of one year and not susceptible of being subsequently called to the colors for reservist training.

On March 7, Mr. Lloyd George submitted an amendment to the above prescribing that the enlistments be voluntary and for a period of twelve years, without the right to discharge except for disability.

On March 10, the matter came up for discussion.

Mr. Lloyd George, in defending it, argued that with a one year period of service Germany could create an army of 2,000,000 men in ten years. He further stated that Great Britain would never sign a treaty fraught with such awful dangers to her security.

In rebuttal, Marshal Foch said that, "While it is true training for one year would produce soldiers of sorts, two hundred thousand of them would be far less dangerous than one hundred thousand professionals of the type proposed by Lloyd George."

In sustaining the Marshal, General Degoutte said, "Such a force will make Germany much more formidable than will any number of one year conscripts." Generals Degoutte, Weygand, and Cavallero then entered a formal protest against allowing the 200,000 twelve year professionals. The compromise resulting from this argument produced the clause in the treaty fixing the German Army at 100,000 men.

Note. The above facts were secured from the stenographic reports taken at the time, and were made available through the courtesy of the State Department. Peculiar significance attaches to this fear of professionals when it is remembered that the men voicing it were all leaders of conscript armies in a successful war.

Opinions of leaders

Von Seeckt

Ten years later, Von Seeckt in his "Armies of Today" says, "When recourse must be had to arms, is it necessary that whole peoples hurl themselves at each other's throats? Can masses be handled with decisive strategy? Will not future wars of masses again end in stalemate? Perhaps the principle of the levee in masses is out of date? It becomes immobile; cannot maneuver. Therefore it cannot conquer; it can only stifle." And again, "The levee in masses failed to annihilate decisively the enemy on the battlefield. It degenerated into the attrition of trench warfare. Germany was beaten down; not conquered. The results of the war were not proportionate to the sacrifices."

Of course, it may be urged that Von Seeckt is simply making the best of a bad bargain; but, is he?

Debeney

At least his former enemies take him seriously, as witness the following: "Germany has in effect 250,000 regulars of long service. We are prone to believe that this is the best modern form. This is human nature, for in general the conceptions of armies oscillate between two poles; the Nation in Arms and the Professional Army." (the military security of France, general Debeney, 1930.)

To meet this menace he is inclined to think that France should have an equal number of professionals immediately ready on the eastern border as the covering army. By giving the men homes in the garrison towns they would be content. Note the similarity with the Burges of Charlemagne.

TERGE

General Terge in "The Protection of Our Frontiers" (also published in 1930) says, "The professional army has these advantages; quality over quantity; instant readiness for war; ideal for offensive." He notes that the Act of 1928 which reduced the term of service to one year, may eventuate either in a return to a longer period of service or in a professional army. He says that in 1914 France and Germany had nearly similar armies; since then they have followed opposite roads; France towards Militia, Germany towards Professionals. Exactly the opposite to the situation in 1870.

As a solution to this menace he suggests the placing on the eastern frontier of a covering army of regulars equal to the German Army, backed with prepared works and machines. Behind this concentration, the militia army.

IV. CHARACTERISTICS OF MASS ARMIES

GENERAL

Admitting the cogency of the historical examples quoted as showing the human tendency to oscillate between extremes, it is still desirable to try to find out what are those characteristics of mass armies which cause some of their recent and most illustrious commanders to view them askance.

In seeking the answer to this question, the writer has asked many officers, including students and instructors at the Army War College, what in their opinion made mass armies desirable? The majority had never considered the case. Such armies existed and, hence, were to be used. "They were the rule," so to speak. Others based their advocacy on one of two reasons:

First; the enemy would have them. Second; that, due to political expediency, they were the only type we could get. Neither reply is convincing. Later, we shall point out the advantage of being different from the enemy. As to expediency, victory is also expedient, and the type of army most likely to secure it will be used.

ADVANTAGES

In general, it seems that the advantages pertaining to the use of large conscript armies are as follows:

First: the sense of power and consequent security aroused in the popular mind by an armed force numbered in millions.

Second: the opportunity to arouse popular enthusiasm and, hence, popular support by placing the burden of war on all alike.

Third: the opportunity of producing homogeneity by a maximum use of local recruitment.

Fourth: the safeguard afforded to political leaders in that if things go wrong they can say that everything possible to secure success had been done.

Fifth: the belief that a national army is the cheapest form of national security.

Sixth: the fact that in a non-likely situation, so far as the United States is concerned, of fighting several major enemies at one time, an army of millions would be demanded to furnish defenders for the several battle fronts.

Seventh: finally, the belief that the expression "big battalions" and "strong battalions" are synonymous.

LIMITATIONS

MEANS OF COMMUNICATION, GEOGRAPHICAL.

The adequacy and number of roads, railroads, and navigable rivers put a definite limit on the size of armies.

Where the means necessary to move the supplies to feed and maintain masses do not exist, masses cannot be used.

This being so, the suitability, or rather usability of masses in different theaters can largely be determined in advance.

DISCUSSION OF ROADS AND RAILROADS IN POSSIBLE THEATERS OF WAR.

In the theaters of the World War where really large forces were employed, 100% of the roads were improved.

Now it is a fact that, in order to maintain the armies occupying these theaters, the roads were used to their maximum capacity; while the splendid network of strategic railroads and the small size of the theaters of operations made the hauls comparatively short.

Further, it seems certain that had the Air Forces been sufficiently powerful to prevent any considerable amount of the daylight movements indulged in, even these roads would have been inadequate. Unless all signs fail, the air forces of the next war will be able to prevent all movement of supplies by day from taking place in the zone of the armies.

Again, it is noteworthy that in this country there are 26% of improved roads, a parenthetical figure of 6%, showing the percent of surfaced roads, is more important because, for military supply in all weathers, only the surfaced roads are useful since shallow ditches and inferior surfaces used in America let unsurfaced roads become waterlogged.

In Europe, on the other hand, this is not the case because, due to ages of tamping and deep drainage ditches, the improved but unsurfaced roads have all-weather usability.

We find that in the United States the elimination of unsurfaced roads makes the area jump from 4.5 square miles to 16.7 square miles, whereas in western Europe there is one mile of improved road to each 0.6 square mile.

The number of square miles served by one mile of railroad is striking since here again the great superiority of western Europe is demonstrated.

DISCUSSION OF ROAD AND RAILROAD DENSITY IN CRITICAL AREAS OF THE UNITED STATES.

An examination of the above mentioned comments may give rise to the thought that the comparisons obtained are unfair due to the many sparsely inhabited areas in all parts of the world, save western Europe.

The average road density in our northeastern area is one mile of surfaced (i.e., improved) road to each one and eight-tenths square miles. In western Europe, as has been stated, the density is one mile of improved road to each six-tenths of a square mile. On the west coast the comparison is even less favorable.

From the foregoing it is evident that the size of armies and the density of the road and railroad nets must vary in an exact ratio in so far as large forces are concerned.

Applying this fact to our own case we find that there is only one place in the world (western Europe) where we can possibly use armies of the size and organization contemplated in our General Mobilization Plan. In all other theaters such masses will either starve or be reduced to impotency.

This fact, taken in conjunction with the admittedly greater mobility and fighting value of small, highly trained and lightly equipped armies, is a weighty argument for our giving serious consideration to the organizing of our forces along lines of more general usability.

Before leaving this subject it is well to ponder the statement made by General Ragueneau, Q.M.G. of the French armies. In speaking of the possibility of pursuing the Germans to the Rhine, he says in effect, that if all the motor transport of the French Army, including the vast pool usually reserved for troop movements, had been used wholly for supply (the railroads being out) it would have been impossible to maintain more than half of the army and that only up to a distance of fifty miles from the railhead. Remember he had three times as many roads as exist elsewhere in the most favored sectors.

EXPENSE

While the staggering expense of one hundred and eighty-six billion dollars chargeable to the World War is only partly attributable to the numbers engaged, the vast postwar costs under which this country is now laboring are the direct result of the size of the forces involved.

Pensions and bonuses are fixed institutions with us. To include June 30, 1931, our veterans have cost us fourteen billion five hundred and twenty-nine million dollars (Veterans Bureau). This staggering cost is due not only to the huge numbers involved, but also to the voting power which large organized minorities possess and use in securing legislation favorable to themselves. With smaller forces both the actual cost and the political power would be much less.

COMPLEXITY OF EQUIPMENT AND LACK OF TRAINING

The more complex the weapons of war become, the less efficiently are they used by partially trained troops. Too often wars are discussed from the standpoint of materiel, rather than from the standpoint of men. Perfect men are assumed. As a matter of fact, only prolonged habit can induce nervous and exhausted men to perform the simplest tasks automatically under fire. When confronted with the manipulation of complex weapons and the use of intricate tactics, the efficiency of both the men and the weapons dwindles towards zero.

While national pride makes us reluctant to admit defects, common sense forces concurrence with du picq when he says,

> Troops will rarely fight unless forced to do so by discipline. Two hundred thousand men, only half of whom fight, are not nearly so effective as one hundred thousand, all of whom do. Those who don't fight still get hurt and all must be fed. It is time we understood the lack of power in mob armies. Time is necessary to give the officers the habit of command, the men the habit of obedience. Victory is to the strong, not the big battalions. Sixty men who can beat a thousand are the stronger. Such men are not numerous. Gideon got only three hundred out of thirty thousand.

INDECISIVENESS

The indecisiveness of the World War was not directly due to the numbers involved, but to the form of combat evolved to use those numbers.

"The advent of huge masses incapable of rapid maneuver made it necessary to rest the flanks on natural obstacles. Also, long fronts were necessary in order to employ the strength of the masses." (Bernhardi's "war of the future.") This, of course, is due to the fact that beyond certain limits depth of formation does not permit rear elements to be usefully employed.

This evolution of the so called "Linear Strategy" was a completely new departure. Prior to 1914 the chief aim of every commander had been to defeat his enemy by maneuver, but when flanks disappeared so did maneuver; and war had to be carried on solely through the medium of frontal attacks which had for their object either the tactical discomfiture of the enemy by attrition, or else the creation of a set of false flanks by penetration.

But masses have other drawbacks besides their inability to maneuver. The areas they occupy are so large that all strategic movements must be made either by train or by truck. Due to this fact the probability of detection from the air is increased and the direction of movement limited to that permitted by the existing roads and railways.

Overhead is excessive. Because of the quantities of supplies consumed, large armies can only exist in areas having (as has already been pointed out) adequate roads and railroads. To operate these means of supply, quantities of laborers are necessary. Not only are the supplies large in themselves, but the less well trained the troops are the more technical and fire assistance do they need, and consequently the number of guns and the amount of ammunition, increases out of all proportion to that needful for the support of trained troops. (Von Seeckt.)

Professor R. M. Johnston puts this very aptly when he says in his "Reflections on the Campaign of 1918," "Low training and high powered equipment tend to produce long (indecisive) wars."

Since mass armies depend on inertia for defense and attrition for attack their use of military stratagems is limited to one field; namely, that of rapid concealed concentrations.

Further, this volume of men and materiel to be moved is such that this operation can only be achieved in industrial areas.

"In other less populous places small numbers and high quality are needful, for if millions can be handled in northern France tens of thousands may well be excessive in the valley of the Dvina, or of the Saint Lawrence." (R. M. Johnston.)

MOBILITY

Mobility, as General W. D. Connor has said, "Is the most abused word in the military lexicon." Mass armies are inherently the negation of mobility. Compared to small professional armies their speed and ability to move are almost zero. Yet, while the static effect of mass on the last war is well known, we still hear it repeated with "parrot like" unction that "Modern war is a question of machines." As a matter of fact, "Science, which devises the machines, works both ways; hence it is wrong to speak of the triumph of the machine over man. The machine has defeated masses of men, not man himself. It never will, as it only comes to life in the hands of man."

"The mistake comes from exposing immobile, nearly defenseless masses of humanity to machines. The greater the masses the surer the victory of the machines." (Because not only is the mass hopeless, but the larger it is the less well is it armed and the less well can it be trained.)

"As science places more inventions at the disposal of the army, the greater are the demands on the soldiers who use these inventions." (Von Seeckt.)

The last paragraph of this interesting quotation is of particular interest in the argument it presents in support of the proposition that a recurrence of small professional armies is imminent. Unquestionably, tremendous combat results could be attained by the use of specialized machines and weapons, manned by men trained and habituated in their use.

OBSOLETE EQUIPMENT

Invention and improvements are a continuing process, while the manufacture of numerous new arms is a costly one. Hence it is apparent that continually to rearm mass armies with the latest weapons is financially impossible. Therefore, tomorrow, as yesterday, nations trusting to masses are bound to send their young manhood to battle with obsolete or obsolescent weapons, whose effectiveness is still further reduced by inadequate training.

So long as masses fought against masses this condition equalized itself; but when professionals were encountered, even when not armed with superior weapons, the results were disastrous. As illustrative, the following quotations from "Liaison," by General Spears, are interesting; "The 5th Division, engaged on a front of six miles, had together with the 19th Brigade and the Cavalry Division, completely held up von Kluck's attack. Two battalions and a battery in flank guard had, with the help of the cavalry, held up a whole German Corps." (August 24, 1914.)

"On August 26, General Smith-Dorrien, with the Second Corps and the Cavalry Division, checked a vastly superior enemy and then withdrew in good order in broad daylight."

"This well nigh miraculous result was in great part due to the fact that the British (regulars) had established themselves as such formidable fighters that the enemy simply did not dare to tackle them save with the utmost precaution."

On the other hand, it is economically and physically possible to keep constantly re-equipping a relatively small force of professionals and at the same time thoroughly train them in the tactics and technique of their weapons. Such a force, well led, could defeat a vastly superior national levee.

Apropos of this is a statement by Professor R. M. Johnston; "It is not sufficiently realized that the armies that fought on the Western Front were all armies of low training the armies, though they differed much in training qualities, were made up of conscripts and not professional soldiers."

(In 1918) "A force of 100,000 trained professional troops could have marched through any place on the Western Front and in either direction. With such (trained) soldiers formations could be made almost indefinitely thin and flexible."

General Lanrezac, commanding the Fifth French Army in 1914, makes the same complaint about his men as compared with the better trained Germans.

In commenting on this, General Spears says, "The errors and mistakes of the French in 1914, when in spite of great gallantry and fearful losses they were so often unsuccessful, were attributed largely to faulty training of the troops and a complete misconception of the conditions of modern war on the part of the officers. A conscript army will invariably be inferior in this respect to a professional one."

THE EFFECT OF AVIATION AND CHEMICALS

Finally, it is desirable to consider the effect which aviation using bombs and chemicals may have on a war of masses. Giving full credit to the great strides made in antiaircraft defense and to the inevitable interference which one air force will cause the other, it is still almost certain that the great troop concentrations, dumps and base installations of the World War are impossible of future duplication. They are too vulnerable to air attack. Where masses are concerned this fact will lead to such dispersion that many of the troops will be unavailable or unsupplied at the moment of need.

It is believed that the foregoing is an unbiased, though necessarily cursory evaluation, of the general characteristics of large conscript armies.

V. CHARACTERISTICS OF PROFESSIONAL ARMIES

GENERAL

In making such an investigation it is inevitable that certain comparisons had to be drawn between such forces and professionals. In order to give a balanced picture, we shall subject the regulars to more specific comment.

MOBILITY, SUPPLY, AND USABILITY

From the standpoint of mobility, supply, and universal adaptability to all theaters of war, the scales are clearly weighted in favor of the professionals.

EXPENSE

While from the aspect of peacetime maintenance the cost of a professional army is high, it sinks into insignificance when compared to the appalling expense due to wars waged by methods of improvisation; for as Marshal Haig said in his final report, "Great Britain was victorious due to her ability to improvise; improvisation is always expensive."

Our present Regular Army of 130,000 officers and enlisted men cost annually in the neighborhood of $261,503,000.

An analogous army of 15,000 officers and 300,000 men, armed and equipped in the latest manner, would cost $500,179,000.

The United States was in the World War for nineteen months. The direct cost of this operation amounted to twenty two billion dollars, about one billion one hundred and fifty-eight million a month. (Ayres.)

In other words, the direct cost of the World War would maintain an army of 315,000 for forty-two years. Moreover, it is a better financial principle to raise money over a term of years than all at once.

Aside from the considerations above and the possibility that such a force would avert a war, there must be counted the secondary bonus expenses, which have already been mentioned as amounting to date to fourteen billion odd dollars, with no end in sight.

Since then the use of national armies tends to produce long and costly wars and longer and more costly expense accounts later, it is clear that any other sort of army capable of correcting these faults is desirable.

STRATEGIC AND TACTICAL CONSIDERATIONS, GENERAL

"New weapons are usually the starting point for new tactics. Hence, in order to avoid unpleasant surprises, it is desirable to envision the future." (Bernhardi.)

The World War produced, or saw used for the first time, many new arms; while the methods of using certain already known weapons changed so radically that they too may be classed as new. This last statement applies in particular to the massed use of guns and machine guns.

Since, however, military men are essentially conservative there is an inevitable lag phase between the advent of new arms and the appearance of new tactics.

This seemingly is what happened in France and what, in large measure, is still going on.

PRESENT ARMY CORPS, STRATEGIC ASPECT

Let us take an army corps, as now prescribed, for an example and see how it conforms to the strategic and tactical conditions it is destined to meet.

Here is a force of some 80,000 men which in one column covers at least a hundred and fifty miles of road. Obviously it should not be on one road, but as pointed out in the discussion on roads, it may have no choice. The rations alone for this force amount to seventeen hundred and eighty tons daily. Even if three roads are available, at least one day is required to deploy for battle. Due to interference by an active enemy on the ground and in the air, the time may conceivably be much longer. Just how great this delay will be depends on the difference in mobility between ourselves and the enemy.

In static warfare this element of time is not so vital. The corps in such a case is simply a sector command composed of shifting divisions which join and leave singly. Moreover, all movement is carried on behind a continuous line which, even in the emergency of a violent attack, has a battle life measurable in days.

In theaters where the terrain is too extensive, or the communication net too meager to admit of our supply a continuous line with the flanks secured by obstacles, maneuver will return. Under such conditions, has a force of eighty thousand men the ability to maneuver while keeping its elements within supporting distance? If we are operating against an enemy equally ponderous the answer seems to be "yes." But, in what will such a maneuver result? Due to slowness, the enemy can always meet us. The "race to the sea," without the sea, will be repeated. Like a race between streetcars, such operations lack conclusiveness because there is no difference in rate of travel.

General Fuller, writing on Cavalry, states, "The student of history will consequently find that only when organization, tactics, and leadership were such as to allow of the mobility of cavalry being rapidly developed from the stability of infantry has war flourished as an art, and when this has not been possible it has degenerated into a dog fight."

His remarks, while true, are not sufficiently inclusive in that they are applied to tactics. The same thing holds in the field of strategy, but with this difference; neither cavalry nor mechanized troops have sufficient combat ability to secure a major victory unaided. Infantry is needed and to be effective it too must have a faster rate than its opponent. Our present organization, due to its size and the amount of its equipment, does not possess this capacity.

TACTICAL CONSIDERATIONS, PRESENT FORCES

Turning now to the realm of tactics, are the formations which limited training imposes on mass armies any more suitable? If so eminent a writer as General Bernhardi can say that, "It is improbable whether infantry can ever again make a successful attack without a predominant artillery," what does he mean? The answer is clear; a tremendous artillery superiority is the only means of getting an indifferent infantry forward.

To insure such a predominance of artillery fire much time is needed, not only for the making of arrangements but also for the accumulation of the guns and ammunition. In the semi-stabilized situations with which we are familiar and on which our own and foreign tactical doctrines are still based, time was available. Equally important were the roads and railways, dumps and installations by and from which those supplies (millions of shells) could be moved and obtained.

Under circumstances where either terrain, roads, or enemy aviation preclude the assemblage of the guns and munitions an impasse is reached and our heavy divisions become impotent.

REASONS FOR USING MASS SYSTEM

Why then do we cling to the mass system and to the tactics which they necessitate? Probably for two reason; first, conservatism; second, herd instinct. "Manking admires force, mass, size, winds, mountains, seas. The warrior typifies force and is admired. The adoration of masses in war arises from the same emotion." (Du Picq.) The column is a form of mass and contains all of the disadvantages inherent to it. The basic idea seems to be that the physical impulsion which will push it on will produce shock. But, "There is never any shock."

"The heavy Boeotians, under Epaminondas, tried to break the Spartan lines at Leuctra and Mantinea with a deep column as with the ram of a ship; but the front rank stopped dead on coming into contact with the enemy; the rest of the column communicated to it no impulsion whatever because an impulse never comes from the rear." (Commandant Colin; "transformation of war.")

Two thousand years later the same thing occurred at McDonnald's column at Wagram. Out of 22,000 men two thousand reached the position; as there were 7,000 casualties, thirteen thousand must have skulked. Finally, the graveyards of France bare tragic testimony to still other attempts at shock tactics. Progress and firearms had effected only this; at Leuctra the rear ranks stopped; at Wagram they ran, having first added materially to the "butcher's bill"; on the Marne they could not even run.

The sole useful purpose of depth is to replace losses in the front line, not to push it on. It was so that the Romans used their tri-fold maniples. "Greek tactics, the tactics of columns, sprang from mathematics. The Roman from a knowledge of man." (Du Picq.)

Is there not a striking resemblance between the phalanx and the square formation for infantry used during the World War and still practiced? The answer is "yes."

Even if it is admitted that for head on assaults such tactics were necessary, are they in war of movement? Only a tiny percent of the men engaged and endangered can use their weapons. Man can stand only a certain amount of terror, after that his nerves give out and he is temporarily useless. The deep columns suffer from this. The rear units get some casualties and constantly see, without being able to help, the ghastly ravages of war which cumber the line of advance. Terror mounts on terror and there is not the stimulation of action to hold it back. The history of a Leuctra, a Wagram, or a Neuve Chapelle, is bound to be repeated, little as we like to admit the fact.

Once again, the phalanx is defeated. How shall we constitute our legion to meet the conditions so aptly expressed in a recent letter to the writer from General Fox Conner; "One of the outstanding impressions made on participants by the battlefields of the World War was their apparent vacancy. What this actually meant was that men can no longer show themselves in any considerable bodies on the battlefield. In the next war it will be impossible for large bodies to move without enormous intervals and distance. This will be true even at night. It, therefore, follows that all units must be smaller in order that command may be retained, and in order that the larger units may be assembled on the battlefield within the limits imposed by the time element."

EVALUATION OF ARGUMENTS SO FAR PRESENTED

It is believed that the evidence and arguments thus far presented are sufficient to warrant us in saying that there is a reasonable probability that the next war will be characterized by the use of smaller and better trained armies. If this be so it behooves all soldiers who, in the words of General W. D. Connor, "Are desirous of being prepared to fight the next war instead of the last" to consider the organization and tactics of such armies.

In the following an attempt will be made to initiate such an investigation.

VI. CONSIDERATIONS AFFECTING THE ORGANIZATION OF SMALL UNITS

GENERAL

In order that such units may move and fight, "With enormous intervals and distances," they must be self supporting; that is, they must organically contain all of the elements necessary to wholly independent action. Since, however, the separation of artillery into units of less size than a brigade detracts in some measure from its effect, the infantry of these units must either be so mobile that they can avoid superior enemy artillery or else must be so efficient that they can attack successfully, "Without a predominant artillery support."

It is believed that highly trained professionals, armed with the latest types of rifles, light and heavy machine guns, and utilizing the full range of their small arms while being supported by a limited amount of organic artillery, can very easily cope with much superior numbers of conscripts inevitably less well armed and less apt at the use of weapons and ground. Moreover, it has already been shown that communications prohibit the universal or even frequent employment of huge armies. Also, since terrain is seldom so complacent as to provide natural obstacles on which to rest the flanks of a continuous line, small armies will not have to contend with greatly superior numbers nor will they have to assault prepared positions.

Xenophon says, "Be it agreeable or terrible, the less anything is fore-seen the more does it cause either pleasure or dismay." This is still true.

The phenomenon of both sides using identical tactics and methods of war is not new and usually has had the effect of causing indecisive battles. The Greek and Roman civil wars; the three "Battle" parallel order of the middle ages; the rigid deployments and endless sieges in the days of Gustavus, Turenne, and William of Nassau; Napoleon's remarks that he was always successful until the enemy learned to copy his meth-ods; and, finally, the World War, are illustrative.

Hence, it seems reasonable to say that any nation developing a differ-ent type of army possessed of high mobility and superior individual abil-ity, combined with a new method of tactics, will have a marked advan-tage until the enemy copies.

SURPRISE

Surprise is one of the prime requisites to victory. Broadly speaking, sur-prise may be utilized in respect to; time, place, and method. One of the principal reasons for the bloody indecisiveness of the World War, once stabilization had set in, is that surprise was rendered almost impossible. Due to similarity of tactical doctrines, surprise as to method was not practiced for almost three years. The startling successes attending the three departures from this rule but serve to prove it; namely, the first gas attack; Cambrai and the first tank attack; and the German offensive of March 1918.

Surprise as to place was restricted in western Europe by the absence of flanks and by the rabid adherence to artillery preparations; though when the vogue for blasting tactics ran out, some scope for ingenuity was restored in the way of secrecy in concentrations. Surprise as to the time was ruled out by the necessity of using darkness to cover the grouping of the assault masses in the departure trenches. The invariable, "Stand to" at dawn shows to what a low ebb surprise, as to time, had sunk. While an examination of major strategy is beyond the scope of this paper, it is of interest to note that so far as nations using universal service are con-cerned, the only opportunity for strategic surprise lies in speeding up mobilization and concentration beyond the point the enemy conceives possible.

A return to small armies capable of, and adept at, maneuver will restore the art of surprise; though the increased speed of observation planes may, in a measure, remove the power of effecting complete surprise as to place. Even if both armies are professional, maneuver warfare still permits surprise; but in this case it will result more from daring leadership than from simple logistics.

TRAINING, GENERAL

HIGH QUALITY NEEDED

Obviously, the quality of a small army must be high in order to cope successfully with a larger one. The whole argument in favor of professional armies rests on the fact that they can be trained better.

SITUATION IN 1917–18

We know that in our armies of 1917–18 discipline in rest areas was good and maneuvers fairly well executed. In battle, on the other hand, discipline, tactics, and the use of weapons were far from perfect. These defects arose from hasty and incomplete instruction and resulted in a plague of specialists trained by one set of officers and led in combat by another. The men of a company did not know each other or their officers. Community of interest and mutual confidence bred of long association were absent. The "buddy" idea was a literary fiction.

AVERAGE LENGTH OF TRAINING IN THE WORLD WAR

The average American soldier who fought in France had training as follows;

> 6 months in the United States.
> 2 months in France, behind the line.
> 1 month in a quiet sector.
> Total, 9 months. (Ayres.)

Considering the enthusiasm of war and the one month in a quiet sector, it is safe to say that these nine months were equivalent to fifteen months in peacetime.

TRAINING IN CONSCRIPT ARMIES TODAY

Reports filed with the League of Nations show that at the present there are thirty-six nations using conscript armies. The length of service of the principle ones is as follows;

France	12 Months
Italy	18 "
Japan	24 "
Russia (average)	36 "
Poland	24 "
Belgium (average)	11 "
Czecho-Slovakia	14 "
Jugoslavia	18 "
Average	19 Months

Hence these armies are little better trained and no better equipped than were ours in France. Further, their size makes it certain that they will start the next war with the weapons with which they finished the last. Whereas a small army can be rearmed.

With such limited training, reliance will have to be placed on heavy formations, predominant artillery support, and movement inspired by voluminous orders.

A professional army could easily defeat such forces, despite a considerable difference in numbers.

IMPORTANCE OF TRAINING

The question of the training necessary to give it this advantage is of interest.

Caesar says that in winter he so trained his men that when the campaign opened he had only to show them the enemy in order to conquer.

Of these same troops Gibbon wrote, "A Roman field of maneuver only differed from a Roman battlefield in that on the former there was no bloodshed."

The purpose of discipline and training

The purpose of discipline and training is:

First: To insure obedience and orderly movement.
Second: To produce synthetic courage.
Third: To provide methods of combat.
Fourth: To prevent or delay the breakdown of the first three due
          to the excitement of battle.

In every human being there is a natural reluctance to obey another. The purpose of our so-called disciplinary drills is to break down this reluctance and make obedience automatic. For this purpose we employ drills and manuals of arms similar to those used in 1750, with this difference; in the days of Frederick they were battle formations.

The Romans were noted for their discipline, yet a careful examination of contemporary bas-reliefs fails to show any regularity in the angle at which their arms were carried, even in a triumph. By analogy, had present methods been in vogue the Romans would have sought disciplinary drills by aping Alexander, and Frederick by copying Gustavus.

The question of attaining discipline only by the use of combat exercises is vital, and can be achieved, though parades will suffer. Battle is an orgy of organized disorder. The worst possible preparation for such a situation is one of meticulous order.

No formations or manuals should be taught which are not directly applicable to route marches and to combat.

Courage

There is a natural reluctance to admit that all men are not brave. The more distant a war becomes the braver in retrospect do its soldiers seem; yet the very regard we pay to valor shows that it is rare. If all Greeks or all Americans had been heroes, would Achilles and Sergeant York be so famous? If all armies fought to a finish, would Thermopylae be of deathless memory?

To produce synthetic courage, discipline and training must be carried to the point where they become automatic; habitual. To these ends fear of punishment and certainty of reward must be utilized. The solidarity arising from mutual confidence bred of long and intimate association must be exploited. So, too, must unit and professional pride. To attain these ends proper methods (and years in which to digest ideas taught and develop automatic habit) are vital.

It is possible to teach a man to sit on a horse and to learn the aids in a few days. In a few months he can get on well if nothing happens, but to control a horse in the excitement of a polo game and to subconsciously use the aids takes years.

METHODS OF COMMAND

PRESENT METHOD

The present method of controlling units in action depends on detailed voluminous orders and constant communication, and traces its origin to the need of employing large numbers of ill trained men and inexperienced officers in the intricate and methodical operations resulting from trench warfare. For such a situation it was admirable; probably the only solution.

To attempt to continue such a system in a war of movement, even if fought with armies as now organized, is doomed to failure.

Reports from the front are inaccurate when sent, and old when received. During the interval (often a matter of hours) conditions have changed. The leading troops will already be engaged. When this happens the commander in the rear cannot influence the movements of these troops, because units under fire move only in two directions; forward or backward.

If it is a case calling for the launching of the reserve in units the size of a present division or corps, the lag in time (caused by the writing and receiving of the report, the writing and distribution of the order, and the time needed for the reserve to develop and begin the attack) is so great that the chances of its being useful are negligible; the situation will have changed.

In the case of forces composed of small, mobile, self-contained units the method is still less applicable.

The successful use of such units will depend on giving great initiative to all leaders in actual command of men.

PROPOSED METHOD

Under such circumstances the solution of the command problem would seem to rest in using the system called by the British, "The Nelsonian Method," or by our Navy, the method of "Indoctrinated Initiative."

This system is based on the belief that "the Best is the enemy of the Good"; that a simple, mediocre solution instantly applied is better than a perfect one which is late or complicated.

Among leaders of whatever rank there are three types; 10% genius; 80% average; and 10% fools. The average group is the critical element in battle. It is better to give such men several simple alternative solutions which, by repeated practice, they can independently apply than it is to attempt to think for them via the ever fallible means of signal communications.

To put such a system into practice requires frequent conferences between the leader and his subordinates in which he indoctrinates them with his method of meeting a few general situations. This teaching should then be further emphasized by map maneuvers and, finally, by actual maneuvers until the idea of a simple, spontaneous system of "team play" is developed. Under this method, orders will be brief, simply stating the result desired, the time, and the place.

At first glance such ideas appall those of us who are accustomed to existing methods. Reflection, aided either by participation in combat or else by reading the accounts by junior officers of recent wars of movement, shows that the only revolutionary thing about it consists in substituting fact for fiction. After the deployment, battles are fought on initiative of juniors who carry on without orders.

In making these statements we in no way disagree with Grant when he says, "In every battle there comes a time when both sides are ready to quit. Then victory comes to the leader who has the nerve to make one more push." The initial impulse comes from the leader and is based on his character imparted by telepathic emanations, though he can personally influence one or two units. The actual work, however, is done by the subordinate commanders who, due to the small size of their forces, can exercise the personal influence of example. Large units, be they platoons, brigades, or armies, cannot be so reached.

The men who man tanks and machine guns are no different from those who, in the past, wielded pikes and muskets. "The human heart is the starting point of all matters pertaining to war." (Marshal de saxe.) Inspiration does not come via coded messages, but by visible personality. The history of war is the history of warriors; few in number, mighty in personality. In small professional armies a method of selection can and must be used which will insure such leaders. It is true that they will frequently be killed, but the death of a high ranking officer has great inspirational effects. Their business is to win, not simply to survive.

DEFECT IN SMALL PROFESSIONAL ARMIES

Before admitting the unalloyed advantages of professional armies, it is necessary to point out one very vital defect which has been brought into being by the use of gunpowder.

In the days of hand arms, man fought man, and his life depended on his equipment and his skill with weapons. Until one side broke there was relatively little loss. But when the pursuit started so did the slaughter, because the opponents were so close together that the vanquished had no start in the race and hand weapons were not adapted to delaying action. Hence, good professional armies suffered few losses; as long as they won they kept on improving.

The following illustrations are of interest. In each case the victor is placed first:

Pharsalus	Caesar 25,000	losses 200
	Pompeii 60,000	losses 15,000
Cannae	Carthaginians 50,000	losses 6,000
	Romans 90,000	losses 80,000
Agincourt	British 15,000	losses 1,600
	French 50,000	losses 25,000
Leipsig (Some firearms used)	Gustavus 40,000	losses 3,400
	Tilly 44,000	losses 13,400

Gunpowder changed all this, for against bullets man enters not a duel with his fellows, but a lottery with fate. The larger casualties come before the break and defeat is less deadly because the further the opponents are apart at the decision, the better start the vanquished have, and firearms are better adapted to delays.

Due to these facts victorious armies now often lose more than the vanquished, and when this is not the case the losses are at least more on a parity. The result is that armies run down with the influx of less well trained men.

The following examples are illustrative.

(Note. These and the preceding figures come from "the dictionary of battles," Harbottle.)

Kolin	Austrians	54,000	losses 9,000
	Frederick	34,000	losses 14,000
Kurnersdorf	Austrian Russians	90,000	losses 24,000
	Frederick	40,000	losses 20,000
Wagram	Napoleon	150,000	losses 18,000
	Austrians	140,000	losses 33,000
St. Privat	Germans	232,000	losses 20,000
	French	133,000	losses 18,000
1st Bull Run	Confederates	26,000	losses 1,752
	Federals	30,000	losses 1,492
World War	Allies		losses 4,689,200
	Central Powers		losses 2,750,000

This fact is very important to the present discussion; a professional army must either be so good that it is immediately victorious or else like Pyrrhus suffer defeat through victory.

A SOLUTION

In order to supply battle losses in a professional army, an enlisted reserve of regulars who have served one enlistment must be formed and its members given enough pay to insure that their addresses will be known. Since no nation is justified in hazarding its existence on the chance of defeating the enemy before its own original force is exterminated, it will be necessary to back up a professional army with a number of reserve forces.

The function of these reserves will be to occupy defensive fronts if such exist, man lines of supply, and finally, after they have gained experience in the school of actual war, to provide additional replacements for the professionals.

The National Guard is adequate in numbers and training to fulfill this mission.

NOTHING NEW IN METHOD PROPOSED

As always, there is nothing new in the idea of a two-type army. In the discussion of old armies (section II) frequent cases of identical procedure were pointed out.

NOT SAME AS PRESENT SYSTEM

At first glance it may seem that the idea of a twofold army herein advanced is identical with the time honored practice of the United States.

SUCH A NOTION IS AN ERROR

The plan proposed contemplates a regular army, with units on a war footing, backed by a trained group of replacements, and of such a strength that in the critical initial phases of the war it will either be wholly successful or else so damage the enemy that his final discomfiture can be effected at a minimum cost of men and money.

The proposed plan changes the regular army from a school to a weapon.

VII. DISCUSSION OF TYPE OF PROFESSIONAL ARMY TO BE DEVELOPED

DESIRABLE QUALITIES

In seeking the mold in which to form a professional force for the performance of this mission, we should seek to accentuate the qualities of immediate readiness; fast, inconspicuous and invulnerable movement off the battlefield, combined with relatively rapid mobility and high fighting capacity on it.

Wholly mechanized army

General

In thinking of suitable forms our attention is at first naturally directed to a wholly mechanized army. There are, however, certain drawbacks, among which must be mentioned the following.

Cost

High cost and rapid obsolescence of equipment. These factors will force us to begin the war with wholly inadequate numbers and will result in our finding ourselves faced with an hiatus when the initial stock is exhausted and the new vehicles laid down at the initiation of the war have not materialized.

Not always suitable

The next factor to consider is that there do not now exist either in fact or in imagination machines capable of fulfilling all or even a measurable proportion of the functions demanded of armies.

Always a counter

Finally, we know from historical analogy that all new weapons have like new diseases developed a curative vaccine. In other words, that new weapons have invariably been most effective, mechanically, during the brief period between their appearance and the arrival of counter measures.

Antitank weapons, etc.

At the present time large caliber machine guns, small automatic cannon, the 5,000 F.S. bullet said to be in existence, and, finally, but most important of all, the disappearance of their novelty, have rendered armored fighting vehicles less potent than in 1918.

SOME MECHANIZED UNITS NEEDED

We should have mechanized units of several types so grouped that they can effectively cooperate with existing arms in carrying out the several military functions which long experience has assigned to infantry, cavalry, and artillery.

Due to the steady progress of invention, it is undesirable to build large numbers of any one type of machine. Better results will accrue if small but complete units of each new type devised are built and put into service successively; as experience and ingenuity point the desirability of changes. When war starts manufacture should be concentrated on the latest approved model in each type.

MAJORITY OF ARMY NOT MECHANIZED

The great majority of the professional army should consist of the types approved by the test of time and whose functions, despite changes in equipment, have remained constant since the beginning.

EQUIPMENT

The equipment of this "Muscle Army" should be of the latest and lightest type varieties and limited to those types adapted to general, as opposed to special, operations.

MOTORIZATION

Since this country contains 74.4% of all the motor vehicles in the world, possesses four companies operating truck fleets of over 10,000 machines apiece, twenty three companies with fleets ranging from 9,427 to 1,037, and nineteen companies operating from 957 to 186 trucks (National Automobile Chamber of Commerce, 1931), it is clear that ample mechanical transport is available which, if judiciously used, can materially lighten the amount of battle equipment carried by men and animals; if and only if we limit ourselves to small forces grouped in small units. To apply this system to masses would simply cumber the roads to no purpose.

Even if the exigencies of a campaign force the eventual abandonment of mechanical transport, every pound mile saved will have paid for itself in the heightened condition in which the troops enter such a phase.

It is neither necessary nor expedient to possess or maintain such transport in peace. Maneuvers and combat exercises are confined almost entirely to the battle part of war. So far as transport is concerned, all that is necessary is that from time to time a survey be made showing the location, number, and suitability of existing types of commercial vehicles so that when war comes they and their drivers can be acquired. Small armies do not need much transport; industry would not be paralyzed.

New weapons

Since one of the principal virtues claimed for a professional army is that due to its limited size, it may be continually rearmed with the newest types of weapons, it is evident that initially the auxiliary army should use present equipment, otherwise expenses would prevent development.

Organization of auxiliary army

So far as organization is concerned, the auxiliary army, depending as it must, more on quantity than on quality, should be organized along present lines with units at half present authorized strength, because any attempt to utilize the elastic formations adaptable to professionals is doomed to failure due to lack of training.

Tables, illustrative only

The construction of tables setting out in detail the organization recommended for a professional army is not only beyond the scope of this study, but also beyond the capacity of any one individual. The numbers and armament herein shown are simply illustrative of the principles believed essential.

VIII. ORGANIZATION OF PROFESSIONAL ARMY

General

The guiding principle of organization should be the endeavor to devise means of killing without being killed. The best way to accomplish this is to reduce the number of human targets while at the same time increasing their killing power. If these individuals can be widely separated, a further saving in losses is assured.

The drawback to wide deployment comes from the fact that usually the several steps in the echelons of command are charged with supervising so many men or groups that they either reduce the interval to increase control, or else abandon control to maintain the interval.

Since, "Man engages in combat for the purpose of gaining the victory and not for the purpose of fighting" (du Picq), and since it is impossible to gain victory without fighting, control is necessary. The obvious solution is to reduce the size of the groups forming each echelon. Everything is simple until the question of supporting arms versus mobility obtrudes itself.

INFANTRY UNITS

GENERAL

Groups composed of two machine guns and one cannon probably represent the best means of attaining the maximum fire with the minimum men. Unfortunately, such a group is impossible because it is immobile and has so few men that it cannot exercise that threat to close which wins battle.

This being so, we must seek a solution which, while retaining the firepower of the cannon and machine guns, also possesses adequate personnel, is mobile, easily commanded, and still admits of such wide deployments as will reduce its losses in battle and avoid interruptions on the march. As has been already indicated, infinite deployment is easy if we can disregard human nature. Since this is impossible we must utilize it. Man is a gregarious, vain, and at the moment, a mechanically minded animal. These traits should be exploited by grouping him around a machine where the vanity of his self esteem will make him fight.

THE LIGHT MACHINE GUN

The light machine gun provides an ideal nucleus for such a mechanical minded group. Moreover, being a machine it is less susceptible to the palpitations of fatigue or emotion than is the rifle. To insure the homogeneous mobility of the group, horses or tractors as a means of transportation are not included; the weapon must be manhandled. Four men, relieved of the weight of their rifles and bayonets and working in pairs, are adequate, and at the same time they can carry 150 rounds each on the march and 250 rounds in battle; total, 1,000 rounds. Twenty-five hundred rounds will be more than sufficient for such a gun. To get this extra 1,500 rounds, riflemen must be used. If we ask each man to carry into action an added 100 rounds, we get nineteen men, or twenty with a sergeant, as a complement for each gun.

From the view point of target range accuracy the light machine gun is slightly less efficient than is the heavy water cooled variety, but it goes with the assault echelon where its presence is visible and audible. This the heavy gun cannot do. Moreover, the heavy gun is not wholly abandoned.

The platoon and company

Returning to our group or section and building from it as a basis, we get a platoon of two sections and a company of three platoons.

The battalion

A battalion of three such companies totals 466, and will have at its disposal eighteen machine guns and 359 rifles. This gives considerable combat value without violating the principle of keeping the echelons of command small. Further, it is not cumbered with either animals or tractors. On the other hand, it is not wholly self supporting in that it does not contain artillery; but were artillery attached it would not only reduce the march, it would also fail to utilize the range of that arm. The question will be examined later when we consider frontages.

The brigade

Combining three such battalions with a company of heavy machine guns (animal drawn) and a battalion of field artillery (motorized), we get a composite brigade, which forms the next echelon of command.

The choice of the name "brigade" is open to criticism. It was adopted in recognition of the fact that while separate arms are reluctant to serve under a colonel of any one of them, they are perfectly content if his name is changed to that of "general." In order to avoid overhead and delay in the transmission of orders, either the brigade or the regiment had to go.

In considering this composite brigade it will be noted that the machine gun company consists of four platoons. Three platoons are the .30 caliber gun and mount; the fourth is a platoon of .50 caliber, air cooled machine guns.

The purpose of this company is two fold; first, it can give the usual supporting fire of machine guns; second, in circumstances where enemy tanks are expected, two battalions can be provided with a platoon of excellent antitank weapons. Unquestionably, it would be nice to have more antitank weapons available, but they cost men and road space; so they have been eliminated. The outstanding defect of our present organization arises from the fact that we have yielded too often to this tendency to be prepared for anything and everything, with the result that we cannot move.

Animals have been retained for draft purposes on the grounds that at the speed required they give more fluid mobility than do machines. Moreover, they can be replaced by requisition in any theater of war.

An examination of the formation of the field artillery battalion reveals the fact that it is composed of two batteries of 75mm guns and one battery of 75mm howitzers. (This gun is the present pack artillery weapon mounted on wheels.)

Again, this battalion serves two purposes; it gives to the brigade the normal supporting fire of artillery, but being in the brigade makes that small unit self contained. The battery of howitzers is also available to supply a platoon of accompanying guns or, if on the defensive, antitank guns to two battalions. The range of these weapons is 9,200 yards, which for the supporting purposes in maneuver war is enough.

THE DIVISION

Combining three such brigades with other units we produce the Division.

Little argument is needed to defend the incorporation of a squadron of seven observation airplanes. The number is only half of those now used, because of the reduced size of the division.

The tank battalion of thirty-nine machines of the Vickers-Armstrong or the modified Christie type is added; first to permit a means of further exploiting the mobility of the division on the offensive; and, second, when on the defensive, to provide a means for local counter attacks.

The organization proposed for this battalion is as follows; the platoon consists of three tanks and one self propelled small cannon (about six pounder) mounted on an identical chassis. While admitting the expensive nature of self propelled guns, they none the less are believed necessary because, in open warfare, the platoons will operate on broad fronts and will often meet situations demanding immediate supporting fire; they, too, must be self contained.

The company consists of three platoons and the captain's tank. The battalion of three companies.

Post war reflections on the circumstances attending war of movement have already caused several countries, including England, to incorporate cavalry in their divisions.

The platoon of armored cars with the squadron is for the purpose of bridging the gap between air reconnaissance and the relatively close work performed by the horsemen. In addition to reconnaissance, in general the squadron will also be useful in covering night movements of tanks, in acting as a mobile reserve, and for general security.

CAVALRY UNITS

GENERAL

In preparing an organization for Cavalry, principles adduced for Infantry are equally applicable. The organization herein presented is practically identical with that at present in effect, except that the numbers have been reduced.

CAVALRY PLATOON AND TROOP

In the troop there are three rifle platoons and one light machine gun platoon. Aggregate force; 116 men, 6 light machine guns, 84 rifles.

In a purely dismounted fight the three squads of the machine gun platoon are incorporated one to each rifle platoon, and an organization similar to the infantry, but with less ammunition, is provided.

When cover permits the use of the mobility of the rifle platoons for mounted envelopment, regardless of whether the fighting conducted is mounted or dismounted the light machine gun platoon forms the pivot about which the rifle platoons maneuver.

CAVALRY SQUADRON

The squadron consists of only two troops. (Aggregate; 241 men, 12 light machine guns, 173 rifles.) To give it three would make it too bulky for command.

CAVALRY BRIGADE

Again and for the same reason, the regiment is eliminated and the brigade formed from three squadrons, with the addition of a three platoon machine gun troop, combining the .30 caliber and .50 caliber guns, a troop of armored cars and a battery of horse artillery using 75mm howitzers.

A study of successful cavalry operations in both the Civil War and in Palestine, and of the German Cavalry in Russia and Rumania, clearly demonstrates the advantage of having batteries attached to small mounted units.

The troop of thirteen armored cars is necessary, as in the case with the divisional cavalry, to bridge the gap between the airplanes and the horse patrols for reconnaissance. When roads do not permit the use of these vehicles, they will be held in reserve with the division. Aggregate for brigade; 1,145 men; 36 light machine guns; 8 machine guns, .30 caliber; 4 heavy machine guns, .50 caliber; 562 rifles; 13 armored cars.

CAVALRY DIVISION

The division consists of three brigades with some trains, signal troops, medical units, and a mechanized cavalry regiment.

It will be noted that the cavalry tank squadron of the mechanized regiment is arranged with three tanks and one accompanying gun in the same manner and for the same reason as in the infantry tanks.

Further, it should be noted that the type of machines used for a mechanized regiment should be constructed to carry out the minor combat functions of cavalry, whereas tanks with the infantry division should be constructed for major combat operations.

SUMMARY

It is desirable to repeat here that the organizations specified above are only illustrative. More critical study will probably point to the need of certain changes in the ratios of the several arms. In making such a study the tendency to provide for everything should be discouraged; that way lies immobility.

CORPS AND ARMIES

It is believed that since the organization of corps and larger units depends wholly on the theater of war, it is foolish to specify them in advance. All that is necessary is to provide the suitable special arms, such as the air forces, medium artillery, engineers, etc., which they will have to use.

However, in view of the probable difficulties of moving large units, as brought out by General Fox Connor, these corps troops will usually be attached to divisions; the corps commander acting in fact as the commander of the theater of operations.

IX. TACTICS OF PROFESSIONAL ARMY

GENERAL

The chief defects in the present methods of attack result from the fact that the masses used are clumsy and intricate; hence hard to deploy and impossible to control. The columnar formations of attack emphasize these defects, while at the same time they subject a great many men to danger while producing an assault echelon lacking in fighting power. Under this system the only means of victory seems to be to collect quantities of guns and mountains of ammunition and then, by having masses of infantry, to hope that some will survive to reap the victory blasted for them by the guns. An enemy capable of more rapid movement will either not select ground suitable for such tactics or else will not wait calmly while the preparations, days in the making, are perfected.

INFANTRY TACTICS, OFFENSIVE

GENERAL

If, then, we hope to conquer by other means than having his infantry scavenge in the wake of the shells, we must get some power into our firing line. If we adopt the present methods of a deployed line of skirmishers, we get little power at the start and none at the finish. The obedience secured from short service men is mental, not automatic. When such men are deployed their mental reactions are dimmed by excitement and a sense of isolation; many do not press on when ordered, because being isolated and invisible, fear of being thought afraid does not affect them. This hesitation occurs more at the long and medium ranges than at the short range. There things are so tense that men must move; it is safer to go on than to lie still. "A retreat forward" takes place.

THE PLATOON

The infantry platoon of forty-three men suggested differs little in size from the present one of some fifty odd. Hence, if we string it out in a skirmish line the only advantage it would have would result from longer training and its light machine guns; this is not enough.

To overcome the straggling incident to skirmish lines it is proposed to deploy the platoon in line of section columns of files or twos, with the light machine guns leading. The interval between columns will be from fifteen to thirty yards.

The platoon commander will be in the interval and abreast of the guns; he can, therefore, insure progression. Moreover, the crowd spirit and the lure of the machine will be helpful.

To advance, the corporal of one gun squad moves out and finds a new location to which the gun is moved under protection of the other. The second gun moves up. Then the platoon commander. Finally, the platoon sergeant and the two section sergeants see that the columns advance. (Ammunition is handed to the guns starting with the men farthest away. Gun crews reserve theirs to the last.)

When decisive range is reached the riflemen will be deployed.

It is believed that up to close range two machine guns so used will produce much more fire effect than is possible with rifles and auto-rifles, many of which are absent. The rifle grenadiers will be used when necessary to prepare the assault.

While for brevity the case has been described with the guns and riflemen staying together, there is no reason why the sections may not be split, or why the gun squads may not stay together and the riflemen be used separately as in the cavalry. Such variations will be made at the shorter ranges.

THE COMPANY

The second platoon of the company, similarly arranged, will be deployed to a flank with an interval up to a maximum of 200 yards. The third platoon, frequently in support, will be in the interval. Its guns should be used whenever cover permits.

THE BATTALION

The interval between companies can be as great as 300 yards.

With two companies in the firing line and one in reserve a battalion will cover a front of possibly 1,000 yards.

THE BRIGADE

The brigade with two battalions in action, with a maximum frontage of 2,000 yards, can still support its whole line with the heavy machine guns and artillery when the latter are held centrally.

It would seem that within limits there will be almost no restrictions to the deployment interval between brigades. Each is self contained and each with one battalion in reserve has a means at hand of checking counter attacks.

This elasticity should be used to cause the enemy either to over extend his inelastic front or else, if he cannot or will not move, to go around him.

THE DIVISION

For turning movements or flank attacks advantage should be taken, whenever the nature of the ground and the presence of cover permits, of using the tank battalion as a means of still further increasing the encirclement.

This is the operation which mechanization enthusiasts dilate on, but they, in imagination, start the move too far from the objective, and then to insure surprise have to assume impossible speeds.

Here we simply move the tanks up under cover behind the most extended flank element of the infantry. No speed is necessary and the ground over which the tanks march is protected. Further, the cavalry will cover the approach of the tanks when they pass beyond the infantry and will reconnoiter roads for them up to the takeoff line. To use tanks for assault over open country is suicide.

In view of the great intervals between all units the need for "indoctrinated initiative and simple, short orders" is highly emphasized.

DEFENSIVE

The teaching as to very wide intervals will still hold. The firepower and mobility of the reserve units will be exploited and, finally, the tanks afford a fine weapon for limited counter attacks. In fact, due to the time and opportunity for careful reconnaissance, they are better for this than for any other operation. But this type of army can be pounded to pieces if it sits down in a narrow country with its flanks on natural obstacles and waits for a blasting attack. Its mobility should enable it to avert such a fate by maneuver. In the cases where special circumstances prevent maneuver the reserve army, as has already been pointed out, should be used to hold the defensive line. While this is going on, the professional army should be employed for flank attacks, breakthroughs or, if the enemy penetrates, for counter attacks.

This does not mean that even in western Europe a mobile army would be impotent. At the start of a war it is impossible to deploy an adequate force behind the whole length of a frontier and then start it rolling in one long, flankless, resistless, wave.

In the first place, the supply difficulties would not permit it; but disregarding this fact, the numbers necessary for such an operation demand a national levee. During the time needed for this and for its deployment a professional army could do a great deal of harm. It might be decisive; if not, it would cause its opponent to occupy very unfavorable ground and would insure ample time for the raising of an opposing national levee to meet him.

CAVALRY TACTICS

The higher mobility given to cavalry by the presence of its horses simply accentuates the capacity for wide deployment as outlined for infantry.

This fact, coupled with the high firepower with which the proposed organization endows it and the advantage which it will frequently derive from the assistance of its mechanized elements, makes it, for maneuver warfare, more effective than it has been for a thousand years.

Nothing in the organization proposed affects the well understood roles of the arms. But the means of accomplishing these tasks have been made more effective.

X. POINTS TO BE EMPHASIZED

In conclusion, the following points should be emphasized:

1. In the World War the use of mass armies produced by nations in arms failed to attain decisive results.
2. Many nations concur in this belief and are seeking a remedy.
3. Historical analogy and enlightened opinion both point to the probability that this remedy will take the form of a war of movement conducted with small, mobile armies.
4. Due to lack of training and natural inertia large national armies are incapable of mobility.
5. Modern war conditions prohibit the movement or control of masses on the march or in battle.
6. Small professional armies composed of smaller self contained units offer a solution for the restoring of mobility and, hence, for shorter and more decisive wars.

SIX

Refining the Concept of Mechanized War

Predicting the impact of new technologies and weapons is an exceedingly difficult task for military thinkers. While it is clear that military innovations can have a profound impact on the conduct of war, ideas about how best to exploit these technological advances often lag behind the acquisition of new platforms. Without a clear vision of how to incorporate new systems and weapons into the force, innovations may create as many problems as they solve by passing increasing uncertainty and costs to their users. With any new technology, it is critical to ask serious questions about its purpose, liabilities, and risks prior to full-scale implementation.

One of Patton's most notable works on the subject of armored innovation was his 1933 article "Mechanized Forces: A Lecture," which was published in the *Cavalry Journal*. The future of tanks, their design philosophy and doctrine, was one of the biggest uncertainties during the interwar period. Although tanks and motorized vehicles made their first appearance during World War I, they were not a fully mature technology and thus had a minimal impact on the outcome of the conflict. The immaturity of the technology and doctrine of armored warfare led to a wide range of theories during the interwar period. The disagreements between competing visions of the future led each nation to adopt different tank design philosophies and doctrines, which in turn produced uneven results during World War II. Patton followed these debates and weapons developments with keen interest and published a series of articles outlining his thoughts on armored warfare.

Because of his unmatched experience as a pioneer of mobile warfare, Patton was uniquely qualified to write on this topic. During the 1916 Punitive Expedition in Mexico, he used a Dodge touring car to lead the first mechanized assault in American history. While on a scouting mission, his unit surprised a group of Mexican bandits who were meeting at a ranch house and killed Villa's deputy, Julio Cárdenas, and two of his men. Patton strapped the dead men to the hood of his car and quickly returned to base before Mexican reinforcements could arrive. This action made international news, boosting the reputation of Lieutenant Patton and earning him, via General Pershing, the nickname "Bandito." In addition to introducing Patton to a broader audience, the action served as proof of the value of vehicles in reconnaissance and raids.

Patton entered World War I as an aide to General Pershing but quickly grew bored with staff work and longed for action. Although he initially wanted an infantry command, he volunteered for the U.S. Tank Corps because he believed the armored branch promised the best opportunity for action and an independent command. Now a tanker, Patton worked with his trademark tenacity to make the new Tank Corps a success. He founded the first U.S. tank school at Champlieu, France, toured the Renault tank factory to learn about manufacturing, interviewed British and French crews to learn from their experiences, mastered the arts of tank driving and maintenance, and became an outspoken advocate for the role of armored vehicles on the battlefield.

In August 1918, Patton took command of the 1st Tank Brigade and led them in the first major tank assault in U.S. history at the Battle of Saint-Mihiel. Patton wanted this first test of battle to reflect well on him and the Tank Corps, and he insisted that his officers lead the attack from the front of the formations and that his tanks never surrender. The attack was a limited success and set the stage for a much larger employment of tanks in September during the Meuse-Argonne Offensive. During this battle, Patton was severely wounded while leading his tanks on foot. He eventually recovered and was awarded the Distinguished Service Cross and Purple Heart for his actions, but the war ended before Patton could return to action. His brief time in combat proved both his personal bravery and the potential value of armored forces.

After the war, Patton was assigned to Camp Meade, Maryland, and was tasked with developing tank doctrine and technology. With the assistance of a young Dwight Eisenhower, he helped write a new manual for tank operations and met with tank designer Walter Christie to discuss the requirements for the next generation of American fighting vehicles. During this period, Patton continued to develop his own views about the proper role of armor in future battles. He became increasingly convinced that tanks were an essential part of modern war and that the United States should create a permanent and independent tank corps. Despite his best efforts, Patton could not get the institutional support or budgetary commitment from the Army to make his vision a reality. He quickly became frustrated with the bureaucratic inertia and transferred back into the cavalry in 1920. Even after his return to the cavalry, Patton never abandoned his belief in the power of tanks and continued to advocate for their development throughout the interwar period.

In this impressive article, which was adapted from a series of lectures, Patton claims that there is much to learn from history about armored vehicles. He expands upon his view that military history is characterized by moves and counter-moves and that tank technology must be continually improved to keep pace with other technical and tactical developments. Given the long development cycles of tanks, Patton was particularly concerned with the obsolescence of American tanks during this period. He also advocated for greater communication and coordination between tanks and their supporting units, asserting that they were essential elements of success in combined-arms warfare. In particular, he placed a high value on radios in vehicles so that troops could make timely reports to base and integrate the battle space as thoroughly as possible.

In the article, Patton argues that tanks' greatest value is that they can restore maneuver warfare and decisiveness to the battlefield. This is a major claim because Patton believed that the skillful use of speed and movement could eliminate the wasteful attrition seen in World War I and allow for decisive action. Patton then expands on a point he made previously, in his War College thesis, that mass conscript armies will no longer be able to hold a continuous and unbroken front and that smaller forces, enabled by machines, will dominate the battlefields of the future.

Despite Patton's generally clear vision of the past and future of armored warfare, in some places this article seems anachronistic. Most notably, he appears to push a weak argument when he insists that horse cavalry will retain a major role in future wars. This argument is especially odd given his personal experiences as a commander of, and advocate for, armored forces. While this may partially reflect the limited technology of the period, it is also likely that Patton's branch loyalty to the horse cavalry and the fact that he was publishing this work in the *Cavalry Journal* influenced his writing. He appears to have purposely downplayed his views about the ascendency of armored forces so as not to upset higher-level officers and entrenched bureaucratic interests in his own cavalry branch.[1] This creates an obvious tension in the article but should not distract from many of Patton's other claims and predictions.

This work again demonstrates Patton's ability to combine historical and practical knowledge. Patton used this ability to synthesize his theoretical and practical understanding to make bold and visionary claims about the future of warfare. While Patton is not considered a prophet of mechanized warfare, as are Heinz Guderian, B. H. Liddell-Hart, J. F. C. Fuller, and Mikhail Tukhachevsky, he was a bold thinker who correctly predicted the nature of and key variables for success in the next war.[2]

MECHANIZED FORCES: A LECTURE

Cavalry Journal, September–October 1933

Many soldiers are led to faulty ideas of war by knowing too much about too little. A picture without a background is both uninteresting and misleading. Hence, in order to paint you an intelligent picture of Mechanization as it exists today, we must provide an historical background.

The appearance of armored fighting vehicles in the World War was a striking reaffirmation of the old adage: "There is nothing new under the sun." After the failure of the German attacks of August and September, 1914, first political and then tactical considerations arose, which made the resumption of a successful offensive well nigh impossible. Neither valor nor ballistics could overcome for long the heightened power of resistance inherent in automatic weapons, barbed wire, and trenches. This ascendancy of the defense over the offense was not new. All through history victory has oscillated between the spear and the shield, the wall and the charge, tactics and technique.

Because of their truly startling parallelism let us investigate two sets of cases. In 1096 B.C., nine years of Hellenic valor had failed to breach the Trojan walls. Then came the Wooden Horse, which by carrying men unscathed within that impregnable circle destroyed in a night Priam's mighty fort. Again in 318 B.C., the walls and ditches of Tyre withstood for a year the furious assaults of the best troops of the day only to fall in their turn before the moving towers of Alexander.

Now let us turn to 1914–16 A.D. Here we find that the inverted wall (the trench) and the inverted ditch (barbed wire) had again rendered

assaults abortive until in their turn they succumbed to the modern version of the wooden horse and the moving tower, which during the winter
of 1915–16 had been simultaneously re-evolved by England and France.
The striking circumstance that, thousands of years later, necessity had
again begat of invention identical solutions for identical problems is
truly arresting.

The French, following the lead of Ulysses, thought of their "chars
d'assaut" as armored carriers destined to transport groups of infantry,
unscathed, across No Man's Land, through the wire and over the trenches
and then disgorge them in the enemy's rear. The British, on the other
hand, followed the Macedonian idea and constructed not carriers, but
mechanical fighters whose duty it was to shoot down resistance, smash
wire, and bridge trenches so as to render the infantry assault less impossible.

Unfortunately for the French plan, that mutual esteem and confidence usually existing between allies prevented either nation from
informing the other of its invention so that, when the French had some
hundreds of machines almost ready for a surprise attack, the British
spilled the beans by jumping off on the Somme on September 15, 1916,
with a handful of tanks. Since surprise, on which the French had counted
for success, was then impossible, they had to revamp their carriers into
improvised fighters. The results of this change were the ponderous St.
Chaumonds and the feeble Schneiders, in which many valiant Frenchmen were roasted and from which few Germans were killed.

The British idea having triumphed, the Allies and later the Germans
made more and more tanks, but, due to the lag phase of about a year
which has always intervened between design and production, the tanks
were always just inadequate to the complete accomplishment of their
tasks. The Mark VIII or, as we call it, the Liberty was the crowning glory
of this lag business, in that, while much money and effort were expended
on it for the specific purpose of forcing the Hindenburg Line, the war
was over some months before the first tank appeared. It is pertinent to
remark that for the future a similar fate probably awaits machines.

As the war progressed a doctrine for the use of tanks was evolved
which was officially stated as follows: "Tanks are an auxiliary arm whose

mission it is to facilitate the advance of the assault infantry. To do this they must so act as to bridge the gap between the lifting of the barrage and the arrival of the bayonet." Towards the very close of the war a corollary was added to the effect that, since machine guns were the enemy to tactical maneuver and tanks were the enemy to the machine gun, tanks had the added function of restoring maneuver to tactics. Within its limits the tank achieved the results as indicated above.

After the Armistice, the natural antipathy aroused in the public mind by the appalling losses of a war of attrition, coupled with the belief that their reduced and dwindling manpower and horsepower would prove inadequate to another such struggle, caused the British to expand the idea of mechanization to the field of strategy, in the hope that by its use they could restore movement and so pave the way for shorter and more decisive wars. While other nations have failed to visualize identical means they are all more or less alive to the necessity of devising some form of warfare which will prevent stabilization. For example, we find General von Seeckt writing, "When recourse must be had to arms, is it necessary that whole peoples hurl themselves at each other's throats? Can masses be handled with decisive strategy? Will not future wars of masses again end in stalemate? Perhaps the principle of the 'levee en masse' is out of date? It becomes immobile; cannot maneuver. Therefore it cannot conquer; it can only stifle." Elsewhere he says, "The 'levee en masse' failed to annihilate decisively the enemy on the battlefield. It generated into the attrition of trench warfare. Germany was beaten down, not conquered. The results of the war were not proportionate to the sacrifices."

Writing in 1930, General Debeney says, "Germany has in effect 250,000 regulars of long service. We are prone to believe that this is the best modern form." As a reason for this statement he says that small armies of regulars are always ready for war and can maneuver fast.

With the possible exception of England most of the thought expended on solving the problem of avoiding stabilization has been concentrated on a solution for the situation as it exists in western Europe. No notice has been taken of the fact that in practically every other possible theater of war, physical conditions exist which of themselves preclude stabilization. For example, in Western Europe, there is one mile of improved hard surfaced road for every six-tenths of a square mile of country. In the Northeastern United States, the next best roaded area, there is one mile of improved road for every one and eight-tenths square miles, only one third as good. For the United States as a whole, the ratio is one to four and a half. In Mexico we find one to five hundred and thirty; in China one to one hundred and twenty-three.

Now we know that in order to maintain the man density necessary for stabilization, even on the relatively short battle front of Western Europe, we used the roads to their maximum capacity. Without pressing the discussion further it is therefore evident that, in bigger theaters of war with poorer road nets, the masses necessary for the holding of continuous lines cannot be supplied and hence cannot be used. Where continuous lines are not occupied, flanks reappear and bring with them their natural corollary, maneuver. In spite of this fact the want of perspective I have alluded to still induces most of us to visualize future battles as simple repetitions of the butting matches of the World War, while soldiers who talk of forces smaller than groups of armies are considered pikers. However, within the last few years, certain signs have appeared which indicate that the tide has turned and that some thought will henceforth be given to fighting wars of maneuver. Let me explain my personal views as to the way mechanized forces will be employed in such wars. We will start with an approved War Department Definition, "A Mechanized Force is one which is not only transported in motor vehicles, but also fights from some or all of them, the vehicles themselves having armament and protective armor." Further, the War Department has decided that the allotment of fighting vehicles to arms shall be along functional lines. That is, vehicles appropriate the traditional tactics of cavalry shall pertain to the cavalry, those appropriate to the traditional functions of the infantry to the infantry, and so on.

Due to the fact that we entered the World War in the middle, we had no experience of those secondary but none the less vital operations incident to the opening phases of all wars and to the entire duration of those waged on the maneuver basis. Since cavalry is the army chiefly used in these so-called minor operations, I shall begin by discussing it and shall point out my conception of how mechanized and horse cavalry will function in such operations.

The chief advantages of Mechanized units are:

1. They possess, under many conditions of terrain and weather, a wider range of strategic and tactical speeds than do any other ground troops.
2. They possess, again under suitable conditions, more rapid tactical mobility than do any other ground troops.
3. Their armor gives them such immunity to many present types of small arms fire that they can develop a maximum of tactical effect in a minimum of time.

Their principal disadvantages are:

1. Being blind, deaf, and having no sensory nerves nor instinct of self preservation, they are very fatiguing to operate.
2. At night, in the presence of the enemy, they are practically incapable of independent movement.
3. They are extremely sensitive to ground and weather conditions.
4. They are no longer a novelty.
5. The increased use of large caliber antitank machine guns and the reported invention of a 5,000 foot per second .30 caliber bullet will increase machine casualties.

Remembering these things let us see how we may employ machines in minor operations. Heretofore such tasks as reconnaissance, counter reconnaissance, the seizure of critical points, delaying actions, flanking operations, and the combats incident to the same have devolved on the cavalry and the air corps.

For the purpose of strategic reconnaissance the armored car occupies a position intermediate between the airplane and a horse patrol. When terrain and weather permit, armored cars can go far and fast; they can secure both positive and negative information and obtain identifications. Their radio equipment should permit them to make prompt reports. On the other hand, their inabilities at night limit their employment.

Armored cars can locate the critical points on the contour of the enemy advance when such points occur on the roads but they cannot trace the curve between the highways nor can they maintain continuous observation. Hence, when the enemy is distant their observations are adequate; as he draws nearer and more minute information is important, they need help.

As the opposing forces approach each other, both sides will attempt to veil their movements by the use of counter reconnaissance. It will then be necessary to fight for information. In 1914, the British stated that all the information they got had to be fought for. Where the resistance encountered is of a minor nature, armored cars can brush it aside. Where it is more serious or where the country is wooded, full of tall crops, or mountainous, the cars lack the necessary combat power and must be helped. The form in which this assistance should be supplied depends on the distance to the front at which the contacts occur. If close in, horse cavalry is best; if farther out, light tanks, or as they are called in the cavalry, combat cars, will be needed. Moving on roads already patrolled by the armored cars, the tanks can go faster than horses and for a longer time. When they arrive they have sufficient cross country power to make limited turning movements and so compel the enemy to either pull out or show his strength.

For distant reconnaissance against a determined enemy and for pursuits, still another type of mechanical unit is necessary.

Any stream large enough to be shown on a one-inch map is an obstacle to machines. If it is defended it is a serious obstacle. Many motor maniacs do not admit this, but talk largely of using their speed to go around. When, however, we consider the difficulty of getting orders to mechanized units, the time necessary to determine and then reconnoiter new routes, and the delays incident to enemy actions, it is certain that mechanized units must often choose between forcing a passage or abandoning a mission.

To force a passage a bridge head must be established; to do this we must have footmen and in considerable numbers. If these men are transported in trucks much time is lost in detrucking on the road, often at the limit of artillery range, and then deploying into approach formation and walking to the firing line while carrying their accompanying weapons. For a force which must depend for success on celerity such a procedure is too slow. To be available in time, these foot fighters or portee troops must be conveyed in light unarmored track-laying vehicles which can move across country when that country is covered by the armored cars and tanks. Moving fan-wise, these carriers deploy under cover close to the scene of action, and their crews (less the driver) have only a short walk into combat.

Before leaving the question of mechanical reconnaissance, it is useful to point out that in horse cavalry we have at all times the three types of units so far described. Patrols equal armored cars, mounted reserves equal tanks, and dismounted troopers equal foot fighters. As ever, there is nothing new. Only the speed ranges and the universality of employment differ somewhat. Next, it is interesting to recall that in war the maps are of small scale, signs missing or in a foreign language, and the people often hostile and always dumb. Try driving at forty miles an hour in a strange country without signs and see where you get. Finally, let me remind you that since, for the immediate future at least, the major parts of all armies will be muscle propelled, information of conditions miles in advance will often be stale before those needing it arrive.

A British writer states that had mechanized forces existed in Palestine and Mesopotamia in 1917–18, the greatest distance to the front at which they could have been usefully employed would have been 150 miles. Beyond that range the number of supply trains doubles, and intermediate camps must be established.

For counter reconnaissance, armored cars are adequate on the roads by daylight. Off the roads, or anywhere at night, neither they nor tanks are useful. Without lights they are stationary; with lights they can be avoided. A fair sort of screen could be made by establishing a line of standing patrols from men in the portee echelon. However, better results will come from using horse cavalry for counter reconnaissance and backing it up with the mechanized forces as a fast reserve to move rapidly to any point where a penetration threatens. You will please notice that, since the horse cavalry covers the front, the mechanized force is immune from the need of reconnoitering for itself, so can go fast. Where columns of machines must move without previous reconnaissance, their rate is very slow as they can be so easily ambushed.

All operations incident to the seizure of critical points, delays, flanking operations, and pursuits demand for their successful accomplishment rapid reconnaissance, fast marching, short violent attacks, and the holding of delaying positions. A command consisting of armored cars, tanks, and foot fighters carried in track laying vehicles possesses all the elements save one necessary to the accomplishment of the above tasks, either alone or in conjunction with horse cavalry. The missing element is, of course, supporting artillery.

On the offensive a mechanized force such as just described would work in general as follows; cover its defensive flank with armored car patrols, dismount some of its portee elements supported by the attached artillery to execute the holding attack, send the rest of the portee elements and all the tanks by road preceded by the armored cars as advance guard to some place from which this maneuvering force can launch an attack against the enemy's flank or rear. When the attack starts, the armored cars, relieved of advance guard duty, assume the role of flank patrols. Here we have the tanks as the charging element, the portee troops as the dismounted cavalry, and the armored cars as patrols.

On the defensive, the foot fighters, deployed at very wide intervals, hold the line; great extension is permissible as the carriers are deployed behind the line like lead horses and no employment is necessary in withdrawing, as is the case where infantry have to converge on trucks. The artillery supports the line. The armored cars cover the flanks, and the tanks act as a mounted reserve.

Thus far I have confined my remarks chiefly to machines acting alone, as this is the most novel and least well understood problem now confronting us. It is my opinion, however, that such operations will be the exception rather than the rule and that in general, mechanized and horse cavalry will operate together. When the two types are combined we have nothing complicated to distract us, since both possess identical tactical and strategic characteristics, the relative advantage being the ability to shift from one to the other according to the nature of the terrain in which the actions occur.

Very often it will be necessary to form composite commands in which combat cars and carrier units operate directly with horse cavalry. Think, for example, of the possibilities of a combat car charge instantly exploited by horsemen. Or of a pivot of maneuver formed by portee troops, while the combat cars and horsemen move out rapidly to clinch the victory by a flank attack.

For night marches, and there will be many of them in the next war, machines must always be preceded by horsemen or else become the victims of ambush.

Coming now to major operations and still remembering the functional distinction of which I have spoken, we find that machines used in major operations act as infantry and belong to it. In offensive battle, it is my opinion that tanks should be held as an offensive reserve for the delivery of the main blow. The timely employment of a reserve composed of footmen in a force the size of a division is most difficult due to the lag which exists between the moment when the situation indicates its use and the time it gets into action. In the corps the conditions are even worse.

Geographically, the area occupied by a tank unit is much smaller than that occupied by an equivalent force of infantry. Hence, the tanks are easier to hide and can come closer to the front.

Tanks move at least four times as fast as infantry.

Tanks develop the full power of their blow at once, infantry must build up its attack.

When tanks are used in this way their assault must be prepared by the greatest possible artillery concentration. If an air attack using bombs and smoke can just precede the tanks, so much the better. Tanks need all the help they can get. Antitank weapons are improving daily, and the novelty which saved us in France no longer exists.

On the defensive, infantry tanks and cavalry mechanized forces will be used for offensive returns against enemy enveloping movements or for direct counter attacks against penetrations.

The portee units of mechanized cavalry will also be very useful in filling temporary gaps in a line of battle, though horse cavalry is generally more suitable, since it is even less a slave to roads.

Possibly some of you may have noticed that so far I have not dealt with the famous American pastime of raids. A moment's reflection should convince any one that the advent of the radio and the airplane have made this always dubious operation still less promising. Secrecy, night marches, the ability to live off the country, avoid roads, and swim rivers are more important than ever. Mechanized forces have none of these qualities. The operations of large independent mechanized forces much heralded abroad are nothing but big raids and are discarded for the same reason.

Next it is pertinent to consider the question of where the machines we talk about are coming from. At the moment, the United States possesses some old Renault tanks and some Mark VIIIs. While neither make has any of the characteristics of a modern fighting machine, as hoped for, except armor plate, they will be used in an emergency, at least they will draw fire.

Of the few machines built since the World War, only about one half have armor plate. This procurement of such plate is most difficult, and this fact will materially limit the speed of hasty rearmament.

Certain writers have said that just as the Mongols conquered by exploiting their resources in horses and horsemanship, so should modern industrial nations conquer by exploiting their supremacy in the automotive world. The comparison is not exact. The Mongol used in unaltered form his normal means of transportation and food the horse. Had some abstruse military reason made it necessary for him to fight only on, "Gray Mares with one China eye," his style would have been cramped, his numbers reduced, and his replacement problems augmented. Armored fighting vehicles are Gray Mares. They are special costly machines with no commercial use. Hardly a part of them is standard. Also, they become obsolescent before they are finished. For this reason no nation will ever start a war with many machines. Those that exist will be expended rather rapidly. Suppose we put the date of their final extinction at three months. Those who know state that a period of from twelve to fifteen months will elapse before replacement machines planned to be manufactured at the beginning of the war will become available. This means that, for a period of from nine months to a year, mechanized forces will cease to exist except for some extemporized armored cars on commercial chassis. Yet, fighting will still go on. God takes care of horse replacements.

In closing, let me remind you of just one more thing. When Samson took the fresh jawbone of an ass and slew a thousand men therewith, he probably started such a vogue for the weapon, especially among the Philistines, that for years no prudent donkey dared to bray. Yet, despite its initial popularity, it was discarded and now appears only as a barrage instrument for acrimonious debate.

History is replete with countless other instances of military implements each in its day heralded as the last word, the key to victory, yet, each in its turn subsiding to its useful, but inconspicuous niche.

Today machines hold the place formerly occupied by the jawbone, the elephant, armor, the long bow, gun powder, and latterly, the submarine.

They, too, shall pass. To me it seems that any person who would scrap the old age-tried arms for this new "ism" is as foolish as the poor man who, on seeing an overcoat, pawned his shirt and pants to buy it.

New weapons are useful in that they add to the repertoire of killing, but, be they tank or tomahawk, weapons are only weapons after all. Wars may be fought with weapons, but they are won by men. It is the spirit of the men who follow and of the man who leads that gains the victory. In biblical times this spirit was ascribed and, probably with some justice, to the Lord. It was the spirit of the Lord, courage, that came mightily upon Samson at Lehi which gained the victory, not the jawbone of an ass.

SEVEN

Training the Force

Preparing an untested force for war is one of the most difficult and important tasks a military can perform.[1] The last article Patton ever published, "Desert Training Corps," addresses this subject. This article appeared in the September–October 1942 issue of *Cavalry Journal* as Patton was preparing to lead his largely untested troops into battle in North Africa as part of the Operation Torch landings. In it, his bold style and talent for self-promotion are clearly evident as he reflects on his founding of the Desert Training Center and his contributions in preparing an untested group of soldiers for the coming campaigns.

Throughout his career, Patton was an outspoken advocate for the value of realistic training.[2] From his very first posting at Fort Sheridan, Illinois, to the creation of the American tank school in World War I and the rigorous training of the Third Army prior to its landings in France, Patton made training a priority.[3] He believed that realistic training would pay dividends by strengthening the physical constitution of his men, creating trust and unit integrity, and allowing units to test their skills in a non-life-threatening environment. These beliefs were summarized in his often repeated aphorism, "A pint of sweat will save a gallon of blood."[4]

As the United States began to prepare itself for entry into World War II, Patton was provided with several opportunities to test both his abilities and the readiness of his men. In a series of war games known as the Tennessee, Louisiana, and Carolina Maneuvers, Patton observed and commanded armored units in realistic large-scale operations. As he participated in these exercises, he solidified his reputation as a bold commander and helped demonstrate the value of concentrated armored forces in achieving rapid victory. In the Carolina Maneuvers, Patton was particularly bold, driving his armored units far around the flanks of the opposing force and capturing his old rival, Gen. Hugh Drum. His boldness impressed the VIPs in attendance, including Gen. George Marshall, who noted Patton's performance with great satisfaction.

Although the maneuvers' participants made numerous mistakes, the maneuvers had value, according to General Marshall, because "mistakes [were] made in Louisiana, not over in Europe."[5] Although the United States had not yet entered World War II, Patton and the Army learned crucial lessons about the need for training that more closely resembled modern combat, the difficulty of coordinating between different combat arms, the value of rapid movement, and the critical role that fuel supplies play in sustaining modern armies. For Patton, these exercises validated his views about the need for rapid movement as well as the increased role that armored forces would play in the upcoming conflict.

Once war was declared, Patton's first task was to help prepare the Army for combat. He was given command of the 1st Armored Corps and tasked with creating the Desert Training Center. Patton was provided considerable freedom of action in establishing the new school, and he personally scouted locations and designed the curriculum for training. From day one, the new outpost reflected the personal touch of its commanding officer. The area he selected for the center was a desolate stretch of the California desert that he remembered from his youth. The curriculum was designed to rapidly toughen up his men while instructing them in the rudiments of desert survival and fighting, and Patton himself observed training at the center from the King's Throne, a large hill overlooking the main training area. He had a radio installed on this observation point so that he could quickly lavish praise or scorn on his men.[6] In his short tenure as commanding officer of the Desert Training Center, Patton prepared over 60,000 troops for battle and established a regimen of drill and discipline that would influence countless others. Although his tenure as commander of this training facility would quickly be overshadowed by his combat record, it should not be ignored.

Patton's goals in writing this article were to praise the hard work and dedication of his own troops and argue for the value of training for the Army as a whole. While this article is not as explicitly rooted in historical examples as his previous works, it clearly reflects a lifetime of reflection on the nature of war. The enduring lesson of warfare is that whereas tactics and technology are in a continual state of flux, people are always the key to victory. To maximize the talents and sacrifice of these people, it is essential that they are properly prepared for the task at hand.

In this essay, Patton abandons his previous claims that horse cavalry should play a key role in future wars but continues to insist that human factors will retain their primacy. This represents both a maturity in the state of mechanized technology as well as an evolution in his thinking. While Patton maintains his insistence that human factors will decide future battles, he stresses that even the best men must be properly trained with modern weapons and tactics. For Patton, this realistic training environment is essential for maximizing the fighting power of his troops, as evidenced by his time as commander of the Desert Training Corps and his participation in the prewar maneuvers.

In many ways, this was a fitting final article for Patton to publish. When it was released, he was preparing for the Allied landings in North Africa, where he would begin to establish his reputation as one of the finest commanders of World

War II. For the remainder of his career, Patton lived by the principles he espoused in this article. While he was unable to complete his memoirs before his death, this short article contains the basic recipe for Patton's success: a lifetime of study, relentless drive, recognition that war is an intensely human endeavor, and a belief that leadership makes the difference between success and failure.

DESERT TRAINING CORPS

Cavalry Journal, October 1942

To all who for years have been bedeviled by arbitrary restrictions on maneuvers, the situation at the Desert Training Center is truly as inspiring as it is unusual. In the whole 12,000,000 odd acres the only restrictions as to movement are those imposed by nature. Even so, these are more accurately deterrents rather than restrictions, for, with time and perspiration, you can go anywhere.

Another point about desert training that is alluring, particularly to artillery men, is the fact that one can open fire with live ammunition or drop bombs at any time and in any direction without endangering anyone. The mountains form the backstops and the parapets. As illustrative of this, seven target ranges, two moving target ranges, two mechanized combat ranges, and a normal infantry combat range have been constructed at a total cost to the government of less than one thousand dollars.

Those people who visualize the desert as a flat expanse of glistening sand are in for a rude awakening, for while there are ample pieces of perfectly flat desert, there are other places with rocks, mountains, and trees. In fact, while in some places one is as visible as a fly on a kitchen table, in other places there is sufficient vegetation to conceal an armored corps. There is, however, one striking difference between the cover provided in Louisiana or the Carolinas and the cover provided by the desert—the desert does not include mosquitoes.

Another point of interest is the fact that even in open places where the sparse vegetation does not exceed two and a half feet in height, a whole combat team of armored vehicles and trucks can be so arranged as to be practically invisible from the air at possible altitudes. By this is meant that at 2,000 feet or over, as many as three or four hundred vehicles cannot be picked up from the air if they are not moving. On the other hand, it has been found possible to pick up as small a unit as six trucks at thirty miles from 6,000 feet, when the trucks were moving.

The tactical mission of the force at the Desert Training Center has been to devise formations for marching and fighting which, while affording control and concentrated firepower, at the same time do not present lucrative air targets. It is felt that these ends have been accomplished. Formations now in use can move across country, followed by the combat train, and without halting can deploy into the attack formation and execute an attack, and at no time present any target worthy of bombardment.

There has been developed, also, a method of going into bivouac which is believed to insure protection from night attacks and from air bombardment, yet, at the same time permit a rapid formation for combat or for march. From a tactical standpoint, in addition to attempting to avoid damage from the air, the corps has specialized in combat suitable for attack on other armored units. In doing so, it has not been found necessary to deviate in any degree from the manual provided by the War Department and the Armored Force. Special cases of the general situations envisioned by those manuals have simply existed. If the platoon commanders know their duty and carry it out, and if the higher commanders maintain discipline and supply and have a rugged determination to close with the enemy and kill him, the answer to successful combat against armored units has been found.

In all operations in the desert, the water is reduced to one gallon per man per day for all purposes. In addition, the vehicles have one to three gallons of water to place in the radiators. However, there have been strangely few occasions necessitating the addition of water to the radiators of the vehicles.

The one gallon per man has so far been more than adequate, even when we have operated for three days in succession at temperatures reaching 130 degrees in the sun. The temperature in the shade is not mentioned because here there is no shade.

In desert operations it has been insisted that all cooking be individual or by vehicle. For this purpose "C" ration, or sometimes "B" ration is used. Experience has shown that the answer to producing fire quickly and effectively in the desert is to fill an empty tin can with desert sand, gravel, or soil, up to within about an inch of the top. Soak the contents with gasoline and light it. This gives ample heat and is a fire easily controlled and easily put out.

It has been found that the liner for the new infantry helmet makes an ideal tropical headpiece. It is worn by all members of the command. An investigation of some four hundred selected individuals has demonstrated the fact that while the wearing of colored glasses is comfort-inducing, it is not necessary. Competent medical officers have observed that those who have not worn them have shown no detrimental effects.

If constant first echelon and preventative maintenance is carried on, the vehicles do not deteriorate unduly. This is surprising when it is recognized that the vehicles have been used at least three times as much as in any other station known to the writer. It is felt that this lack of mechanical deterioration is due somewhat to the fact that owing to the nature of the ground excessive speeds are impossible.

The general health of the command is remarkably good. The tendency to obesity is distinctly lacking. For instance, Sergeant "Man Mountain" Dean has, it is said, lost sixty pounds but is still quite a figure of a man!

People are apt to think of the desert as a hot, horrid place. Actually, the heat is much less oppressive than the heat at similar times of the year in Georgia or Louisiana. As for training, the situation is ideal. It should be remembered that from October to the end of May the weather in the desert is what babies cry for and old, rich people pay large sums of money to obtain.

An Intimate Glimpse into the Patton Mind

Seven years after her husband's death, Patton's widow, Beatrice Ayer Patton, attempted to provide some insight into the mind of the famous general. In a short article, "A Soldier's Reading," Mrs. Patton discusses how reading and the study of military classics were integral parts of her husband's character and development. At first glance, this article may appear to be little more than a series of personal anecdotes written by a dedicated spouse, but a closer inspection reveals much more. In fact, this article exposes a deeper, more interesting side of Patton that was largely unknown at the time of its publication in 1952: his lifelong dedication to mastering the military profession.

Even today, one of the least appreciated elements of Patton's character is his brilliant mind. Although Patton is better known for his dashing appearance, dynamic leadership style, physical presence, and unusual beliefs in mysticism and reincarnation, his less visible qualities were no less essential to his success. He had a natural genius for leadership and the military arts, but he worked diligently his entire life to improve both his knowledge of the facts and his analytical powers.

Despite early academic struggles, Patton became a dedicated reader and acquired an extensive library of military texts, which he transported to various posts throughout his career. Patton made extensive notes in the margins of these works and often read and reread various passages to internalize their meaning. To further internalize these lessons, he created a series of note cards that he used for quick reference. Patton's study was truly an attempt to "know the enemy."

As the articles in this volume attest, Patton's writings were remarkable. He used writing to formalize his ideas and expose them to a broader audience. Although Patton left many of his writing projects incomplete, over the course of his career, his published articles and official reports shared his insights with a broad set of strategic thinkers and practitioners. Despite his full-time commitments as an active-duty officer, he was able to pursue a level of personal professional study that carefully considered and accurately predicted many of the most important trends in military affairs.

Patton dedicated his life to the study of the military arts, and to this end, he was extremely successful. His efforts serve as an example of how even the most naturally gifted strategists must continually search for insights from the world around them. A final example of Patton's dedication to studying the military arts comes from his campaign in North Africa. During the uneventful passage across the Atlantic prior to Operation Torch, Patton read the Koran, hoping to gain wisdom about the peoples and cultures he was about to encounter. Although Patton was a pious Christian (if unconventional and profane), he greatly enjoyed the text, declaring it "a good book and interesting."[1]

This study paid unexpected dividends as Patton used his knowledge of Islamic religion and culture to successfully co-opt local power brokers and adroitly manage a potentially hostile Arab population as he fought his way across North Africa. This forgotten incident is just one example of how one of Patton's lesser-known characteristics—his mind—was a critical element of his success. For George Patton, success was not an accident.

A SOLDIER'S READING

Armor, November–December 1952

BEATRICE AYER PATTON

It began with the classics, for the Pattons felt that life was too short to get one's education unless one started early, and the family loved to read aloud. By the time the future general had reached age eight, he had heard and acted out *The Illiad, The Odyssey,* some of Shakespeare's historical plays and such books of adventure such as *Scottish Chiefs,* Conan Doyle's *Sir Nigel, The White Company, The Memoirs and Adventures of Brigadier Gerard, The Boys' King Arthur* and the complete works of G. A. Henty.

As a cadet he singled out the great commanders of history for his study, and I have his little notebook filled with military maxims, some signed J. C., some Nap, and some simply G. Sources were his specialty, and as a bride, I remember him handing me a copy of von Treitschke, saying: "Try and make me a workable translation of this. That book of von Bernhardi's, *Germany and the Next War,* is nothing but a digest of this one. I hate digests." Unfortunately, my German is not of that caliber, and he had to make do until a proper translation was published several years later. He was, however, one of the first Americans to own that translation, as later he owned translations of Marx, Lenin and the first edition of *Mein Kampf*—believing that one can only understand Man through his own works and not from what others think he thinks. No matter where we moved, there was never enough room for the books. We were indeed lucky that an Army officer's professional library is transported free.

He made notes on all the important books he read, both in the books themselves and on reference cards, and he was as deeply interested in some of the unsuccessful campaigns, trying to ferret out the secret of their unsuccess, as he was in the successful ones. I have one entire book of notes on the Gallipoli campaign. He was especially interested in landing operations, expecting to make them himself someday.

Our library holds many works on horsemanship, foxhunting, polo and sailing, all sports with a spice of danger to keep a soldier on his toes in time of peace.

He was an intensive student of the Civil War, and one of his regrets was that his favorite military biography of that period was a foreigner . . . Henderson's *Stonewall Jackson*. Imagine his delight when Freeman's *Lee* began to appear. He bought and read them one volume at a time, and when I showed it to the author, crammed with my husband's notes and comments, he smiled: "He REALLY read it, bless his heart." His memory was phenomenal and he could recite entire pages from such widely different sources as the *Book of Common Prayer*, Caesar's *Commentaries* and Kipling's and Macaulay's poems. On the voyage to Africa in 1942, he read the *Koran*, better to understand the Moroccans, and during the Sicilian campaign, he bought and read every book he could find on the history of that island, sending them home to me when he had finished them.

During the campaigns of '44 and '45, he carried with him a Bible, prayer book, Caesar's *Commentaries* and a complete set of Kipling—for relaxation. A minister who interviewed him during that winter remarked that when he saw a Bible on his table, he thought it had been put there to impress the clergy, but had to admit later that the general was better acquainted with what lay between the covers than the minister himself.

Most of all, he was interested in the practical application of his studies to the actual terrain, and as far back as 1913, during the tour at the French Cavalry School, we personally reconnoitered the Normandy Bocage country, using only the watershed roads used in William the Conqueror's time, passable in any weather. When he entered the war four years later, he fought in eastern France, but in 1944, his memory held good. People have asked me how he "guessed" so luckily.

"Terrain is sometimes responsible for final windup of a campaign, as in the life of Hannibal," he wrote. To him, it was not a coincidence that the final German defeat in Africa was near the field of Zama. His letter, "I entered Trier by the same gate Labienus used and I could almost smell the sweat and dust of the marching legions," is an example of how dramatically he could link the present with the past. As he had acted out the death of Ajax on the old home ranch, so he and our family acted out

Bull Run, Chancellorsville and Gettysburg. I have represented every-thing in those battles from artillery horses at Sudslcigh's Ford to LT Cushing, Army of the Potomac, at the battle of Gettysburg. That was a battle long to be remembered. At the end of the third day, as the girls jumped over the stone wall into Harper's Woods, Ruth Ellen fell wounded, took a pencil and paper from her pocket and wrote her dying message. (The original by COL Tazewell Patton, C.S.A., is in the Rich-mond Museum.) I heard a sort of groan behind me. As LT Cushing, firing my last shot from my last gun, I had been too busy to notice a sightseeing bus had drawn up and was watching the tragedy of Pickett's Charge.

If I have digressed from my subject, reading, it is to show the results of reading. First he studied the battles; then, when possible, played them out on the ground in a way no one who ever participated in the game can forget.

From his reading of history, he believed that no defensive action is ever truly successful. He once asked me to look up a successful defensive action . . . any successful one. I found three, but they were all Pyrrhic victories. History seasoned with imagination and applied to the problem in hand was his hobby, and he deplored the fact that it is so little taught in our schools, for he felt that the study of man is Man, and that the pres-ent is built upon the past.

As I read the books coming out of this last war, I know those that he would choose: authoritative biographies and personal memoirs of the writer, whether he be friend or enemy. No digests!

Mrs. Patton's annotated list of General Patton's favorite books

Maxims of Frederick the Great.
Maxims of Napoleon, and all the authoritative military biographies of Napoleon, such as those by Bourrienne and Sloane.
Commentaries, Julius Caesar.
Treatises by von Treitschke, von Clausewitz, von Schlieffen, von Seeckt, Jomini and other Napoleonic writers.
Memoirs of Baron de Marbot of de Fezensec, a colonel under Napoleon: we were translating the latter when he went to war in 1942.
Fifteen Decisive Battles of the World, Creasy.

Charles XII of Sweden, Klingspor.

Decline and Fall of the Roman Empire, Gibbon.

Strategicon, Marcus and Spaulding.

The Prince, Machiavelli.

The Crowd, Le Bon.

Art of War in the Middle Ages, Oman, and other books by him.

The Influence of Sea Power on History, Mahan, and other books by him. (The Trilogy.)

Stonewall Jackson, Henderson.

Memoirs of U. S. Grant, and those of McClellan.

Battles and Leaders of the Civil War; R. E. Lee: A Biography and *Lee's Lieutenants: A Study in Command*, Freeman.

Years of Victory and Years of Endurance, Arthur Bryant.

Gallipoli, Hamilton.

Thucydides' *Military History of Greece*.

Memoirs of Ludendorff, von Hindenburg and Foch.

Genghis Khan, Alexander and other biographies, Harold Lamb.

Alexander, Weigall.

The *Home Book of Verse*, in which he loved the heroic poems.

Anything by Winston Churchill.

Kipling, complete.

Anything by Liddell Hart, with whom he often loved to differ.

Anything by J. F. C. Fuller, especially *Generals, Their Diseases and Cures*. He was so delighted with this that he sent a copy to his superior, a major general. It was never acknowledged. Later he gave 12 copies to friends, colonels only, remarking that prevention is better than cure.

NOTES

Introduction

1. Carlo D'Este, *Patton: A Genius for War* (New York: Harper Collins, 1995), 118; and Roger H. Nye, *The Patton Mind: The Professional Development of an Extraordinary Leader* (Garden City Park, N.Y.: Avery, 1993), 27.
2. Quoted in Nye, *Patton Mind*, 27.
3. Ladislas Farago, *Patton: Ordeal and Triumph* (Yardley, Pa.: Westholme, 2005), 66.
4. For a fascinating book describing the film, its accuracy, and its enduring legacy, see Nicholas Evan Sarantakes, *Making Patton: A Classic Film's Epic Journey to the Silver Screen* (Lawrence: University Press of Kansas, 2012).
5. On Patton's swordsmanship and views on fencing, see George Patton, "Saber Exercise," War Department Document No. 463 (Washington, D.C.: Government Printing Office, 1914).
6. D'Este, *Patton*, 816.
7. Harry H. Semmes, *Portrait of Patton* (New York: Appleton-Century-Crofts, 1955), 42.
8. Martin Blumenson, *Patton: The Man Behind the Legend, 1885–1945* (New York: William Morrow, 1985), 168.
9. D'Este, *Patton*, 699.
10. Alan Axelrod, *Patton: A Biography* (New York: Palgrave Macmillan, 2006), 67.
11. On Patton's skilled handling of the local Arab leaders during the North African Campaign, see J. Furman Daniel III, "Patton as a Counterinsurgent? Lessons from an Unlikely COIN-danista," *Small Wars Journal*, January 25, 2014; and George S. Patton Jr., *War as I Knew It* (Boston: Houghton Mifflin, 1947), 5–39.

12. On this point, see, for example, Alan Axelrod, *Patton's Drive: The Making of America's Greatest General* (Guilford, Conn.: Lyons Press, 2009); and Michael Keene, *Patton: Blood, Guts, and Prayer* (Washington, D.C.: Regnery, 2012).
13. Harry Yeide, *Fighting Patton: George S. Patton Jr. through the Eyes of His Enemies* (Minneapolis: Zenith Press, 2011), 5.
14. On the deficiencies of U.S. professional military education during the period, see Jörg Muth, *Command Culture: Officer Education in the U.S. Army and the German Armed Forces, 1901–1940, and the Consequences for World War II* (Denton: University of North Texas Press, 2011).
15. For a fascinating account of Patton's reading and study of the military arts, see Nye, *Patton Mind*.
16. Ibid., 104–8.

Chapter One. An Early View of a Military Mind

1. D'Este, *Patton*, 138–40.
2. Semmes, *Portrait of Patton*, 29.
3. Patton, "Saber Exercise."

Chapter Two. The Warrior Mindset

1. For an enduring, and still controversial, thesis on the link between the decline of Roman military virtues and the empire's eventual collapse, see Edward Gibbon, *The Decline and Fall of the Roman Empire*, vols. 1–6 (New York: Random House, 2010). For a more modern view on the compatibility of the "millennial" generation and military service, see Matthew Hipple, "Now Hear This—The Millennial Debate Presents False Choices," U.S. Naval Institute *Proceedings* 140 (November 2014): 341.
2. Nye, *Patton Mind*, 76–80.
3. Dennis Showalter, *Patton and Rommel: Men of War in the Twentieth Century* (New York: Berkeley, 2005), 113.

Chapter Three. Technology and War

1. Patton, *War as I Knew It*.
2. Semmes, *Portrait of Patton*, 52–53.

Chapter Four. Patterns of Success

1. Examples of this debate are Gian Gentile, *Wrong Turn: America's Deadly Embrace of Counterinsurgency* (New York: New Press, 2013); and Victor Davis Hanson, *The Savior Generals: How Five Great Commanders Saved Wars that Were Lost—From Ancient Greece to Iraq* (New York: Bloomsbury Press, 2013).
2. D'Este, *Patton*, 699.
3. Nye, *Patton Mind*. See generally Carl von Clausewitz, *On War*, ed. and trans. Michel Howard and Peter Paret (Princeton, N.J.: Princeton University Press, 1976).
4. Victor Davis Hanson, *The Soul of Battle: From Ancient Times to the Present Day: How Three Great Liberators Vanquished Tyranny* (New York: Free Press, 1999).
5. Quoted in Nye, *Patton Mind*, 65–66.

Chapter Five. Anticipating the Next War

1. Edward M. Coffman, *The Regulars: The American Army, 1898–1941* (Cambridge, Mass.: Belknap Press, 2004), 285.
2. Martin Blumenson, "George S. Patton's Student Days at the Army War College," *Parameters* 5, no. 2 (1976): 25–32.
3. Williamson Murray and Allan R. Millett, eds., *Military Innovation in the Interwar Period* (Cambridge: Cambridge University Press, 1996).
4. James S. Corum, *The Roots of Blitzkrieg: Hans Von Seeckt and German Military Reform* (Lawrence: University Press of Kansas, 1992).

Chapter Six. Refining the Concept of Mechanized War

1. George F. Hofmann, *Through Mobility We Conquer: The Mechanization of U.S. Cavalry* (Lexington: University Press of Kentucky, 2006), 178.
2. John J. Mearsheimer, *Liddell Hart and the Weight of History* (Ithaca, N.Y.: Cornell University Press, 1989).

Chapter Seven. Training the Force

1. Charles E. Heller and William A. Stofft, eds., *America's First Battles, 1776–1965* (Lawrence: University Press of Kansas, 1986).

2. Barry D. Watts, *U.S. Combat Training, Operational Art, and Strategic Competence* (Washington, D.C.: Center for Strategic and Budgetary Assessments, 2008).
3. Axelrod, *Patton*, 27.
4. Blumenson, *Patton*, 168.
5. D'Este, *Patton*, 392–407.
6. Ibid., 406–13.

Conclusion

1. Patton, *War as I Knew It*, 5.

FURTHER READING

Axelrod, Alan. *Patton's Drive: The Making of America's Greatest General.* Guilford, Conn.: Lyons Press, 2009.

Barron, Leo. *Patton at the Battle of the Bulge: How the General's Tanks Turned the Tide at Bastogne.* New York: Penguin Group, 2014.

Daniel, III, J. Furman. "Patton as a Counterinsurgent? Lessons from an Unlikely COIN-danista." *Small Wars Journal,* January 25, 2014.

D'Este, Carlo. *Patton: A Genius for War.* New York: Harper Collins, 1995.

Essame, Herbert. *Patton: A Study in Command.* New York: Charles Scribner's Sons, 1974.

Farago, Ladislas. *The Last Days of Patton.* New York: McGraw-Hill, 1981.

———. *Patton: Ordeal and Triumph.* Yardley, Pa.: Westholme, 2005.

Hymel, Kevin M. *Patton's Photographs: War as He Saw It.* Herndon, Va.: Potomac Books, 2006.

Keane, Michael. *Patton: Blood, Guts, and Prayer.* Washington, D.C.: Regnery, 2012.

Nye, Roger H. *The Patton Mind: The Professional Development of an Extraordinary Leader.* Garden City Park, N.Y.: Avery, 1993.

Patton, George S., Jr. *War as I Knew It.* Boston: Houghton Mifflin, 1947.

Patton, Robert H. *The Pattons: A Personal History of an American Family.* New York: Crown, 1994.

Rickard, John. *Advance and Destroy: Patton as Commander in the Bulge.* Lexington: University Press of Kentucky, 2011.

———. *Patton at Bay: The Lorraine Campaign, 1944.* Dulles, Va.: Brassey's, 2004.

Sarantakes, Nicholas Evan. *Making Patton: A Classic Film's Epic Journey to the Silver Screen.* Lawrence: University Press of Kansas, 2012.

Yeide, Harry. *Fighting Patton: George S. Patton Jr. through the Eyes of His Enemies.* Minneapolis: Zenith Press, 2011.

ABOUT THE EDITOR

J. Furman Daniel III is an assistant professor at the College of Security and Intelligence at Embry-Riddle University where he researches and teaches at the nexus of political science, political theory, and military history. His hobbies include book collecting, running, performing amateur archeology, traveling, and spending time with his wife, Christina. He holds a BA (with honors) from the University of Chicago and a PhD from Georgetown University.

The Naval Institute Press is the book-publishing arm of the U.S. Naval Institute, a private, nonprofit, membership society for sea service professionals and others who share an interest in naval and maritime affairs. Established in 1873 at the U.S. Naval Academy in Annapolis, Maryland, where its offices remain today, the Naval Institute has members worldwide.

Members of the Naval Institute support the education programs of the society and receive the influential monthly magazine *Proceedings* or the colorful bimonthly magazine *Naval History* and discounts on fine nautical prints and on ship and aircraft photos. They also have access to the transcripts of the Institute's Oral History Program and get discounted admission to any of the Institute-sponsored seminars offered around the country.

The Naval Institute's book-publishing program, begun in 1898 with basic guides to naval practices, has broadened its scope to include books of more general interest. Now the Naval Institute Press publishes about seventy titles each year, ranging from how-to books on boating and navigation to battle histories, biographies, ship and aircraft guides, and novels. Institute members receive significant discounts on the Press' more than eight hundred books in print.

Full-time students are eligible for special half-price membership rates. Life memberships are also available.

For a free catalog describing Naval Institute Press books currently available, and for further information about joining the U.S. Naval Institute, please write to:

Member Services
U.S. Naval Institute
291 Wood Road
Annapolis, MD 21402-5034
Telephone: (800) 233-8764
Fax: (410) 571-1703
Web address: www.usni.org